Advance

Barbara Harris provides an excellent resource that cuts through old myths and misinformation to give the reader her well-informed considerations for exploring the new world of medicinal marijuana use.

With the increasing number of states and countries legalizing marijuana for medicinal use comes the urgent need for reliable information and guidance on the positive uses of marijuana and the dangers as well.

Barbara writes from her personal and professional experience, generously sharing her great wisdom distilled from a lifetime of experience.

This is a great book for anyone interested in marijuana use, not abuse.

Lawrence Edwards PhD
Author of The Soul's Journey

This book is wonderful! Thank you for having the courage to write it.

I like the fact that you are recommending marijuana not as just an anti-anxiety medication, but as a vehicle to help explore the underlying issues.

Bruce Greyson MD
Emeritus Professor of Psychiatry,
University of Virginia School of Medicine

As a British psychiatrist, I enjoyed reading this new book.

I believe it will become an important bible for clinicians who may prescribe Medicinal Marijuana.

I was impressed by how well you balanced the scientific issues alongside personal reflections and the latest evidence-based clinical approaches to illnesses that can benefit from Medicinal Marijuana.

Ben Sessa MBBS (MD), B Sc, MRC Psych

Consultant Psychiatrist and Psychopharmacology
Senior Research Fellow, Cardiff University Medical School,
Wales, UK

The Secrets *of* Medicinal Marijuana

A Guide for Patients Those Who & Care for Them

Barbara Harris, RT, CMT

Author of
Spiritual Awakenings & The Natural Soul

MHP
muse house press

MUSE HOUSE PRESS

Natural Paradigms

ISBN: 978-1-935827-23-8

Trade Paperback

Copyright 2015 Barbara Harris

All Rights Reserved

Visit us on the Internet at: www.MuseHousePress.com

Muse House Press, the MHP Logo and Natural Paradigms are imprints of Muse House Press.

Cover design and Interior composition by:

Randy Klinger & Esther Klinger / Muse House Press Photo

Illustration by: Kyle Kusa

Direct requests to: design@MuseHousePress.com

Printed in the United States of America

Third printing

Dedication

This book is dedicated to the patient who has never used marijuana and those who have never used medicinal marijuana and the loved ones or caretakers who are assisting them and who may also not know about the secrets of this remarkable plant.

On the Cover

The ancient book appearance on the cover reflects that this plant has been around for longer than we can imagine. There are caveman drawings with marijuana leaves depicted interacting with humans. Eating cannabis as a medicine goes back at least 2,500 years to ancient China and perhaps earlier to treat malaria, rheumatism and menstrual pain. Cannabis use is also reported in traditional Indian medicine. Michael Backes notes that observer Robin Sharma calls it the "penicillin of Ayurvedic medicine." [1]

Appreciation: My deep appreciation goes to Randy Klinger and Esther Klinger the graphic artists who created the beautiful cover design and implementation of the design on each page of this book. Their creativity and enthusiasm were an amazing gift to me that continues to this day.

Thanks to Nancy Strand for helping me envision this book. And to Steve Cartano at CartanoCreative.com for my photograph.

---

[1]   Backes M 2014 *Cannabis Pharmacy*: The practical guide to medical marijuana. Black Dog and Leventhal Publishers, New York, NY p 104

Preface

I neither negate nor recommend 'recreational' marijuana. Instead, in this book I will explore and describe the healing gifts that this unique plant offers to help balance us physically, mentally, emotionally and spiritually.

Used carefully, as I describe in some detail, it has the potential to help us balance these basic parts of our life.

Using it as Medicinal Marijuana (MM) could also raise our level of awareness or consciousness. If you're already familiar with using cannabis, you may know that using it could change your perceptions to include a larger and more inclusive view of yourself, others and the world.

Many users already know that marijuana with high THC content should not be used -

• While operating any vehicle or power machinery.

And what many may not know is that it should also not be used

• While doing anything that needs your full attention and concentration, such as studying in school or doing home work or serious research, when making important decisions or working with your hands on a potentially dangerous project.

This is because using cannabis with high THC has been associated with causing memory defects in both our short term and long term memory and negatively impacting our physical reflexes, movement and balance. While at times using it can trigger new ideas, possibilities and ways of seeing or looking at things, it can also lead us astray in a kind of faux creativity.

Anywhere along the way you are in control. This may include developing and having a better understanding of your personal healing and the way you perceive and understand things about your life, your relationships and the world.

Through the suggestions in this book I will describe how to use MM to show you how you may think at times and how you can change the way you think if you decide to do that.

Each of us is unique with our individual health goals we may want to address by using natural cannabis products. While I am not a physician and cannot give medical advice, I base my information in this book on my own experiences, the feedback I have received from my patients and on many other sources on cannabis information which I cite and reference.

As with any health condition, I encourage you to confer with your physician or healthcare practitioner about your individual problems. The material contained in this book is for informational purposes only. It is not intended as a substitute for the advice and care of your physician or other clinician and you should use proper discretion, in consultation with them, in using the information that I present. The author and the publisher expressly disclaim responsibility for any adverse effects that may result from the use or application of the information contained in this book.

Table of Contents

Figures

Tables

Case Histories

Exercises and Meditations

Text Box Summaries

Appendix Tables

CBD and Some Key Studies

Examples of how early sophisticated studies show some of its healing effects

Cannabidiol (CBD)

- Isolated 1940 (Adams), but identified positively in 1963 (Mechoulam & Shvo)
- Binds CB_1 with Ki 4900 nM and CB_2 4200 nM, but shows unique ability to antagonize these receptors with K_B in low nM range (Thomas 2007)
- Neuroprotective AO, strongly inhibits glutamate excitotoxicity, also anti-oxidant > Vitamins C and E (Hampson et al. 1998)
- Now known to be a VR_1 agonist with EC_{50} 3.2-3.5 μM (Bisogno et al. 2001)
- Inhibits uptake of the AEA, and weakly inhibits its hydrolysis (Bisogno et al. 2001)
- "In a manner of interpretation, CBD may be considered the first clinical agent that modulates endocannabinoid function." (Russo 2003)
- Alerting vs. THC in clinic (Nicholson 2004), and experimentally in rat hypothalamus and dorsal raphe nucleus (Murillo-Rodriguez et al. 2006)
- CBD may have its own endogenous receptor. It is an antagonist at GPR55 and GPR18 (McHugh et al. 2010)

cannabidiol

Russo. E.B. 2011 Taming THC: Potential Cannabis Synergy and Phytocannabinoid-Terpenoid Entourage Effects. British Journal of Pharmacology (in press). Slide courtesy of EBR.

Introduction

What this book is about and how to use it

Over the last 30 years scientists, researchers, plant chemists and herbal farmers have made increasing advances in our understanding of the cannabis plant (marijuana) and how it affects our body, mind and spirit.

Over that time several countries and states have either legalized or decriminalized using it.

While Colorado, Washington, Alaska, DC and Oregon have legalized it, 22 other states have enacted a medicinal marijuana (MM) provision. But in the other 24 states it remains illegal as of June 2015.

And there are *still no clear* or definitive *guidelines* on *how to use* it *as a medicine*.

...This situation is confusing.

A big reason is because cannabis is a plant that contains over 400 chemicals, at least 80 of which are said to be bio-active on our physical, mental, emotional and spiritual well-being.

A meta-analysis of 30 independent published studies on the personal effects of cannabis found that users most frequently reported:

Enhanced relaxation,

Improved mood (i.e., feeling good, content),

Increased insight into self and others, and

Improved perceptions. [2]

Cannabis chemists Arno Hazecamp PhD and George Pappas PhD call for those that have used cannabis in a medicinal way to step forward with their experience and knowledge and help educate the rest of us. They said that this is because establishing medical efficacy through clinical means alone is limited and overlooks multiple potentially useful psychosocial factors. This is why in this book I share my own and my patients' experiences using this remarkable natural plant.

I also describe what many knowledgeable and often savvy scientists and helping professionals have found.

Quality Control

More on the confusion: At present there is no source that makes a reliable specific dose of any cannabis product that physicians or pharmacologists can recommend for a patient to get and take for any specific problem.

---

2   Green B, Kavanagh D, Young (2003) *Being Stoned*: A review of self-reported cannabis effects. Drug & Alcohol Review 22(4), 453-460
(Authors compared self-reported effects in 12 naturalistic and 18 laboratory studies).
See also how Nadia Solowij PhD describes other cannabis effects below.

And: This statement includes the fact that the THC and cannabidiol (CBD) amount, percentage and ratio contained within any cannabis product remain important and at times crucial for the user to know about before taking a dose of it (see also pages 160 and 162 in the Appendix for THC and CBD details). This is because as of today there is no acceptable "quality controlled" source, chemical or dose that we can trust and rely on. Quality controlled drugs are usually guaranteed by the drug maker to have the precise drug at the exact amount and dose, commonly listed in milligrams (mg) on the label.

Today if you get an MM certificate from a state authority, you are on your own to use it safely and effectively. A recent book that can help determine what kind of MM to get and how to take it, including dose, ratio of THC to CBD, and delivery recommendations is *Cannabis Pharmacy*: The Practical Guide to Medical Marijuana by Michael Backes. Other books, articles and websites are available which I show in the footnotes and the references on page 187. I give some examples of dosing throughout this book (see, e.g., CBD oil dosing table on page 124).

In this uncertain climate where several states and countries have suddenly legalized a plant that has been illegal for decades, chemists Drs. Hazecamp and Pappas and countless others who are brave enough, including numerous MDs, are admitting that they don't know enough about how to work with Medicinal Marijuana (MM) but they want to learn more about it.

They urge those of us who have worked with it ourselves and with our patients to come forward. I address that throughout this book.

Waking up to the benefits of MM

This is exciting research because seeing it as I have and watching these articles and reports coming from the scientific paradigm means that we are finally waking up to the benefits of a plant that has been helping mankind for thousands of years.

There is a real difference between cannabis that is smoked recreationally and Medicinal Marijuana. As I will describe throughout this book, it depends on our intention and awareness. And when we become aware of how this intention works and what cannabis/marijuana is capable of, then it and we become capable of helping ourselves to expand our consciousness, to heal our inner life, and may boost our immune system.

In 1860 the French author Charles Beaudelaire said "What hashish (cannabis) gives with one hand, it takes away with the other: that is to say, it gives power of imagination and takes away the ability to profit by it." In this book, I describe what Medicinal Marijuana gives. I will show how to use this power of our imagination and how to profit by applying it in the healing process.

The Australian cognitive psychologist Nadia Solowij PhD wrote, "After euphoria, the acute cognitive effects of cannabis are among those that are most often sought by those who use the drug. The loosening of associations, the intensification of ordinary experiences, a heightening of humor, pleasant imaginative reverie, are all cognitive effects that provide users with a welcome relief from the tedium of everyday life. Some cannabis users experience these effects daily over a period of many years. It has long been suspected that the price paid for the regular elicitation of these diverting cognitive effects of cannabis may be some type of enduring, and possibly irreversible, cognitive impairment, and thus, the loss of ability to profit by its use."

Solowij et al and others have found cognitive (thinking) impairments in long term heavy users, including memory, attention, inhibitory control, executive functions and decision making during the period of acute intoxication and beyond. Throughout this book I will describe how marijuana can help and what it takes away if it is used inappropriately and/or in excess.

The Three Boxes of Using Cannabis

Over time we have slowly developed three ways to look at using cannabis. These three ways include: 1) "Bad" or Illegal, 2) Fun (which we see as often self-medicating for PTSD) and usually called 'Recreational' use, and

3) Medicinal, which helps relieve symptoms or problems. We are beginning to step out of these three boxes that represent the public's perceptions of this plant which I summarize as an illustration of these in Figure 1.

Figure 1 The Three Boxes of Cannabis Use and Medicinal Marijuana

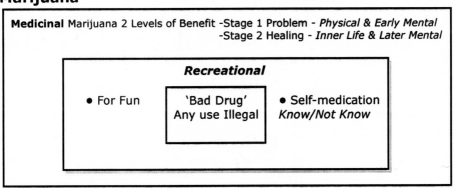

How have these three views come about? Authorities appear to have instituted the first box to increase their power and control of us starting with the 1937 Marijuana Tax Act that over time spawned a massive legal and prison industry (politicians, police, lawyers, courts, jails and prisons—a dynamic clearly shown in a recent documentary *The Culture High* worth watching). These authorities have thereby blocked our ability to experiment on how we might best use cannabis for our benefit beyond what the second box shows. Meanwhile, in the second box recreational users have used it either for fun and/or as self-medication, with most doing the latter unknowingly.

The MM movement expands cannabis use as a legal way to continue experimenting. In this book I will explain several ways to use it directly and by trial and error to help us heal. For example, in Chapter 6 on the Stages of Recovery I give details on how to use it in a Stage 1 Problem - Physical and Early Mental and Stage 2 Healing – in our Inner Life and Later Mental conflicts.

I invite you to open more doors to your own inner life (See Figure 2 on page 57.) An example: some of my cancer patients told me, "I am now grateful for my cancer because of all the gifts it has given me."

Using MM can help us realize these gifts and more that I will describe. I outline opening and using our inner life to bring us into balance because this is a way MM can help us to realize our True Self/Soul in a more real and deeper way.

As you read my own personal story you will realize how closely I have worked with this plant to help myself while I was in a full body cast for seven months. I could then stop taking the opiates and benzodiazepines daily which are toxic and have much worse consequences in their "side effects" (brain disabling, mind numbing, toxicity, and addiction) than Medicinal Marijuana. A few years later when I became a respiratory therapist, I (and many health care providers around me) knew and occasionally recommended Medicinal Marijuana, but we didn't know to call it by that term then.

As I watched my patients with cancer wasting away from the toxic effects of the chemotherapy and the tumors, I quietly took the initiative of informing them on how they could help themselves with cannabis. I couldn't tell them how to get it, but I told them how to use it to help their associated pain. This book contains what I have learned about plants in general and specifically about MM. I want you the reader to know how my story includes learning from an early age and all through my life the gifts that some plants have waiting for us if we are willing to open ourselves to getting to know them and ourselves better.

If you are a patient that has a "terminal" prognosis or you are someone who cares for someone who is going to die, the second chapter is a good one to read out loud to each other because it is the story of an experience that changed my life. It is the near-death experience that I had 40 years ago and it is still with me daily. It is still teaching me about living and it has taken away my fear of dying. I went from being an atheist to knowing eternity in a split second of Earth time.

I learned not to fear death which helped me not fear being alive either. My Life Review showed me that where we live, what we drive and what we wear isn't the primary reason to be here. What life here is all about is being real for myself, others and the God of my understanding.

And with this gift of being real, I realized, and repeatedly heard this as I and my colleagues interviewed hundreds of Near-Death Experiencers, that life is about learning. This physical reality is a school where we spiritual beings are learning how to be human. And that same knowledge can be learned by people who have chronic or acute medical problems. We don't have to nearly die to learn about the true nature of our being here. Medicinal Marijuana, if we wish, if we are open and willing, and if we intend to realize its benefits, is waiting to help us be peaceful and learn. *

> As Dr. John Craven reasoned in his book The Power in Pot: "High" or higher awareness raises the species above a life ruled by fear and gives us a reason for living, for the work of learning and building." [3]

Because Medicinal Marijuana has not been in the public eye until recently, I wrote this book on three levels:

1- for the patient,

2 - for the caregiver and

3 - for the licensed clinician who may not have worked with it before. Thus, I have tried to keep the hardest-to-understand information in the book's rich Appendices. Chapter 11 on cannabinoids may be the most difficult to understand. Research is showing that this part of cannabis has a multitude of healing properties for many conditions and can be taken without experiencing the THC 'High.'

Barbara Harris, RT, CMT Atlanta, Ga, USA
June, 2015, *Updated* January 2016

* And if you or who you care for doesn't wish to go deeper MM is there to help with relaxing, seeing the humor in life and enjoy the munchies. More than that is up to the patient.

---

3   Craven J (2014) *The Power In Pot*: How to Harness the Medicinal Properties of Marijuana ... Kindle edition

1 The Secrets

A secret is usually hiding information from someone, but seldom sharing it. Here I am using the term somewhat differently. I will show eight ways to use Medicinal Marijuana that are mostly unknown to many current users and most non-users.

Nearly all other medicines are man-made chemicals that are supposed to be scientifically shown to reach a presumed goal for our health and then formulated in specific doses and dispensed by a pharmacist, nurse or other professional. Not so for cannabis. It is a plant from nature that has potentially healing properties, but it does not come in specifically measured doses.

I base these secrets on others' and my observations that are both tried and true. They are mostly common sense if you think deeply about it. Gautama Siddhartha, the Buddha, once said "Three things cannot long stay hidden: the sun, the moon and the truth."

In this introductory chapter I will outline each of eight secrets.

I believe that these are the most efficient and effective yet uncommonly known ways to use it. See what you think. Then, in the rest of the book I will describe specific details about how to use each of them.

The Eight Secrets

Use less The first secret is to 1) use less of it, which usually works better. This is in contrast to a 'Cheech and Chong' mentality of the past to use as much as possible.

Cannabis has biphasic properties: small doses and high doses generally have opposite effects. For example, a low dose is commonly anxiolytic (anti-anxiety) vs. a large dose which is often anxiogenic (increases anxiety, even to the point of paranoia or panic).

To get the most benefit, what commonly works best is to use the smallest effective dose possible. The patient thereby takes more effective control of their own dose. This benefit helps us realize that we as patients no longer need to feel passive, since we will now feel and be more in control. We can determine our best dose.

If we consume increasingly more cannabis, we may become numb and get so far away from our peaceful existence in the Now that we lose our ability to function.

Be Now 2) Settle down as our Real Self into the Now. Using it in low doses facilitates our ability to settle down and be and function in the Now, the Present Moment. Within this mental and emotional space we tend to stop worrying about our past and future. Staying in the Now is a way to focus on what is most real for us today.

Being in the Here and Now is both a basic method and goal of meditation and Buddhist/Zen/esoteric Christian/ Taoist/Sufi/

Kabbalistic Jewish spiritual practice as taught by the masters over the millennia and described in some detail over the last century. Attaining this state is reached through various ways Using MM can facilitate this process for some who find it helpful. [4]

Be Aware 3) Use MM with progressively more conscious awareness, which I describe throughout this book. This action includes being fully aware to take responsibility for planning and considering each of the next five secrets.

Be Personal Doing so facilitates 4) using MM to open our own personal door to a natural balance for re-generating our health and healing. This is a private journey. We may risk sharing it with selected safe people who will not judge us for using it, which I address next.

Be safe 5) One of the main cautions for patients is for us to share our experiences and inner life only with people with whom we feel safe.

If we are helping another person, this principle includes not trying to convince them of something they may have shown opposition to doing in the past.

Perhaps in the first MM session we might ask our self:

Who do I feel safe with?

Who do I not feel safe with?

Who do I feel frustrated by?

Might it be better for me to share with safe others only?

---

4   For patients who are suffering/in the process of dying, "occasional" may not be enough. Again, they are in charge of their own dosing and frequency.

4

Address my symptoms 6) start addressing and asking

(a) What are my symptoms or the other person's symptoms ? and
(b) What are the names that best describe these symptoms and what the symptoms may be trying to tell us.

When we name and describe these, we are moving into the new paradigm or medical model that includes Body, Mind and Spirit. We take control of our healing and can investigate more choices to work within an "integrative," "alternative" or "complimentary" medicine approach. The umbrella term for this model is "Holistic Health." Alternative or Integrative medicine physicians may be more accepting to MM use and some may know more about using it than most physicians.

From working in the addictions and trauma focused recovery field for decades, others and I use similar principles for healing that I will explain. The results of these two different but similar paradigms is that instead of casually or thoughtlessly following prescriptions for drugs or "medicine" to cover up our symptoms similar to what 'Band-Aids' do and keep us passive, we learn how to re-balance our own health and well being.

Use Prayer 7) I can use simple prayer to help MM do what I want in healing my symptoms, imbalances or any other problems. Prayer can activate the power of intention to tap into its healing properties. I have two favorite prayers that are easy to memorize. The first is "Help!" The second is "Thank You!" Add anything you would like, but these simple words are often enough.

And remember:

If we are hoping for a miracle
We can make room for it.
Prayer does that.

Take charge 8) We **become proactive** and thereby take charge of our own health care. Using this framework helps us get to the deeper reasons of what is underlying our symptoms and how to work with that knowledge to let go of that deep level so we can remove whatever "imbalance" in our life may be creating or underlying our symptoms.

Carefully using MM can help us find our natural balance or internal stability, what physicians and scientists call *homeostasis*. This is a new way to see our self, others and God. It helps us to get in touch with our Real Self, Soul or the part of us that is eternal. It raises our consciousness and gives us a new perspective on the way things work and a sense of a timeless quality. Here our life slows down and we can get in touch with a deeper part of our inner life. This action includes taking responsibility for assuring a personal quality control to address the ten variables and dosing guidelines described on page 119, and 168 and in the Appendix (see especially Appendix 10 and 11 on pages 175 and 178).

Skip ahead if any of the early reading is not addressing Medicinal Marijuana in the detail you want. Look at the Table of Contents and select what appeals to you. For example, for some readers chapters 3, 7, 8 and 11 may be more like what you want to read first. And the Appendix is rich in MM specific content. I will be adding text boxes that highlight selected scientific reports as examples of key findings or principles, the first of which is on the next page.

US Govt HHS patent on MM   Abstract summary

Cannabinoids as **antioxidants** and **neuroprotectants**

"Cannabinoids have been found to have antioxidant properties, unrelated to NMDA receptor antagonism .* This new found property makes cannabinoids useful in the treatment and prophylaxis of wide variety of oxidation associated diseases, such as ischemic, age-related, inflammatory and autoimmune diseases. The cannabinoids are found to have particular application a s neuroprotectants, for example in limiting

neurological damage following ischemic insults, such as stroke and trauma, or in the treatment of neurodegenerative diseases, such as Alzheimer's disease, Parkinson's disease and HIV dementia. Nonpsychoactive cannabinoids, such as cannabidiol, are particularly advantageous to use because they avoid toxicity that is encountered with psychoactive cannabinoids at high doses useful in the method of the present invention. ..." [Quite a positive observation.]

*the most common brain neurotransmitter glutamate activates the NMDA receptor thereby causing cell damage. "Cannabidiol blocks glutamate toxicity with equal potency regardless of whether the insult is mediated by NMDA, AMPA or kainate receptors." See online at google.com/patents/US6630507 This patent was apparently written in the late 1990s, filed in 1999 and published in 2003. Note online at the bottom the 12 groups or companies who appear interested to develop cannabis into useful treatments and there are some 60 more similar patents in process for a total of 72.

How do you patent a natural plant? We extract legal CBD from hemp that does all of the above and more. Big Pharma may have difficulties finding a single molecule to patent.

Found online in many sites, including leafscience. com/2014/07/25/u-s-government-patent-marijuana/

Part I The Personal Side

"You do not need to know precisely what is happening, or exactly where it is all going.

What you need is to recognize the possibilities and challenges offered by the present moment, and to embrace them with courage, faith, and hope."

Thomas Merton [5]

---

5   Thomas Merton (1915-1968) was an American Catholic writer and mystic. A Trappist monk of the Abbey of Gethsemane, Kentucky, he was a poet, social activist, and student of comparative religion.

Please read this first before beginning Chapter 2

Throughout this book I reflect on my own experiences both with MM and without it.

In this next chapter 2,

I explain the near-death experience and why that experience helped me wake up.

Here I will also give a foundation for the importance of our psycho-spiritual inner life.

Then in chapter 3, as an example of using MM, I detail my use of it to help me through the long healing process from a serious and near catastrophic injury.

If you want to use MM mainly to increase your appetite, decrease worry and stress, lessen withdrawal from opiates, benzodiazepines and other drugs, sleep better and more, you may want to go right to the "how to" clinical section in part 2 on page 39.

2 My Near-Death Experience

In 1973, I injured my lower back when I was pushed unexpectedly into a swimming pool. For two years I was treated "conservatively" which back then meant bed rest with traction and painkillers in combination with muscle relaxants or benzodiazepines.

Going into my third year of debilitation, the doctors and I agreed that I wasn't getting any better. We were just treating the symptoms. I then opted for a spinal fusion operation hoping I could return to a normal life. I couldn't be a mother to my three young children, or a wife anymore. I was desperate and this was the choice I was given.

A two hour operation turned into 5 and a half hours. There was a fracture that hadn't been seen on x-ray. I awoke in a Stryker-frame Circle bed. I couldn't move – the bed moved me by rotating me from my back to my belly so my back could air out (to prevent bed sores) and also drain my lungs. I was to remain like that for a month and then put in a full body cast from my armpits to my groin and then down one leg.

Barbara Harris RT

About two days after surgery I started to die. There was internal bleeding and blood loss. My blood pressure dropped really low and then all my vital signs started to shut down. I remember opening my eyes and seeing this big belly and not knowing what was going on. The pain became so intense that I started screaming. The staff rushed in pushing carts and machinery and talking loud as they were tossing stuff over me. I remember looking up and they were throwing tubes and bottles back and forth and then I realized I must be dying. I was screaming, "No, no, no, leave me alone! Let me die!"

It had been two years of being sick and I just didn't want to live anymore.

First Experience

I lost consciousness and the next thing I knew I woke up in the hallway. It was probably the middle of the night. The lights were dimmed. I looked back and forth and I didn't see anybody. Then it hit me. I wasn't supposed to be out of that strange bed. If they caught me out in the hall I'd get into trouble. So, I turned around to move back into the room and then I realized something was really weird because I was looking into a public address speaker. This PA speaker was mounted on the ceiling and I was looking directly into it. Then I knew that something strange was happening. I moved back into the room and looked down into the circle bed and saw my body.

That didn't bother me. I was peaceful. In fact, I was very calm. It was okay that "she" was down there and I was up where I was. The only thing I remember at that point was chuckling because my nose had tape wrapped around it to hold a tube from moving that went down my throat.

I hung out with "her" for a while. I knew that wasn't me anymore.

I had no problem with this separation and then the next thing I knew I was in total blackness and I remember thinking that my eyes weren't working. Then I felt hands come around the back of me and pull me into this warm soft lush love which I suddenly realized was my grandmother. She had been dead for fourteen years.

When she died I thought that was it! It was over.

I never gave much thought to the fact that when people were dead, they were dead. But I also had absolutely no doubt at that moment that I was with my grandmother. As I realized that I also realized that everything I'd ever believed from before this moment was just a belief and now I was in touch with the real Reality (with a capital R). This new Reality was that my grandmother and I were together again, and as a thought that, every memory we had of each other came alive again and all this love that we have for each other embraced us and was us. And I didn't just see my experience of my memories, I saw it all through her eyes and felt it all through her heart.

I knew she could see and feel my memories too.

I stayed with her for a while and then started moving again through this darkness. As I moved away from her she let me know, heart to heart, that she would always be there waiting for me.

As I looked through this darkness I could see energy churning and the energy was moving/churning through the blackness with light coming out. The light was moving down to the end of wherever I was going and it was forming the Light. I wanted to get to It. I felt the Light. I could hear the slow droning noise that was beckoning me to the Light but at the same time my hands felt like they were expanding. I became fascinated by my hands and the next thing I knew it was morning.

I was back in the circle bed. Two nurses were in my room opening the drapes. I tried to tell them that I left the bed and they told me I had hallucinated. My eyes were now so sensitive to light that I had them keep my drapes closed. My hearing became so acute that we kept my door closed because the voices from the PA speaker sounded shrill.

Second Experience - My Life Review

A week later I left my body again. I had been rotated face-down which was uncomfortable because there was practically no mattress at all. It was just a canvas stretcher coated with vinyl. I weighed only 85 pounds at the time. That is 30 pounds less than I weigh now.

I was skinny and uncomfortable.

The button for the nurse that was usually clipped to the sheet fell away from the bed so I couldn't press the button to call for the nurse, and the door was closed so nobody could hear me calling.

I called and then I started screaming.

I became hysterical.

Suddenly, I separated again from my body and left the bed. This time I was totally awake and I watched it happen. I watched my body move away in the circle bed and I looked out ahead of me and I was back up in the blackness. I looked down and to the right and my body was in that circle bed in a bubble, and she was crying hysterically. I looked out to the left up high into the blackness and there was another bubble, and in it was me at about a year old face down in my crib crying just as intensely. I kept looking back and forth between the two scenes because I was obviously confused, and about the third time I looked back and forth, it hit me that there was a very strong Presence that was an Energy or a Force, certainly not an old man with a long white beard but an Energy that wrapped Itself around me, moved through me and was me. It was pure Love but also unexplainable in this reality.

If you took all the love I felt for and from my grandmother and you multiplied it by a million, maybe then you could get the intensity of this Love. It was a vibratory warmth and It held me up. I realized that as an atheist this couldn't be. That's when I totally relaxed. I fell into this lush softness, this warmth that held me moved into every molecule of my being and let me know that I and everything was It and It was everything too. I totally let go and I somehow knew that I could be letting go of this lifetime too. There was no past, no future. Everything was Now and the Love of this Being holding me. I melted into IT.

The baby in the crib became the center of a cloud of bubbles and in bubble after bubble there was another scene from my life. Together this Amazing Intelligent Being and I bounce through all these bubbles and WE re-experienced my life. We as One re-experienced everything including what the other people in the bubbles experienced too. I wasn't just re-experiencing it as myself but I was my mother and my dad and my brother and my boyfriend who went on to become my husband and then my kids. I was them as well as me.

I could feel our love. I could feel the painful times too.

I could feel everything that was going on

At the same time, I felt this Intelligent Being holding me.

Call IT God, The Universe, All-That-Is. I also called IT "The Force" after I saw the first Star Wars movie. It took me a long time to use the word "God," because there are so many uses of that word and I had experienced this Energy as being beyond any word. As with my grandmother, this Energy let me see my life through ITs eyes and feel my life through ITs heart. And there was no judgment, only love and compassion.

I actually saw and felt my painful childhood and adult life through God's eyes and heart.

I learned that each lifetime in this physical reality is to learn, to grow, to have compassion for ourselves and each other. All of the polarity that exists here doesn't exist where we were in Eternity. We humans invented all of the differences here that separate us from each other. Imagine going into one cell in our own physical body and telling it that it is part of a body of millions of cells.

That one cell won't believe us until somehow someday it can step out of its boundaries and see all of our body. Because of this, the first realization I had was that we're not separate beings. We don't end at our skin. We are all connected.

So, like all the people who have these experiences, we've been zapped into enlightenment for a few seconds out of time and we come back here and don't know what to do with what we were experienced, shown and/or what we learned.

Spiritually Transformative Experiences (STEs)

So while there's enlightenment when we are in our experience and then afterwards, there can be and usually is grieving for having to give up the Light and be back here. Then after this grieving process there is a search to find it again only this time here. Eventually we do but we have to heal ourselves first. (see Chapter 6 on the Stages of Recovery and Holistic Health.)

Unfortunately, in my research I observed many near-death experiencers getting stuck in their newly found psychic gifts that can happen after any spiritual awakening (The experience can be triggered by many other ways than nearly dying. I have listed these and discuss the integration of spiritual experiences in Chapter 12 on page 127.)

Getting stuck in developing our psychic "powers" is a known trap in our Soul's development after a Spiritual Experience. That's why I was and still am focusing on personal growth and not psychic prowess.

Psychic abilities (which I believe we all have with or without a spiritual experience) can be a distraction.

If you are an MM "patient" reading this book, or you are someone who cares for them, be aware that I am describing steps to having our own personal spiritual experience. It all depends on how far and how deep you want to investigate your own True Self/Soul. You can laugh and relax into MM and stop there. You also have the choice of going deeper to eliminate physical and energetic blocks that go deep inside and may be causing the imbalances that are draining and possibly causing your illness.

MM can help us find our own healthy balance, and in that exploration we expand our beliefs and find a new level of expanding consciousness. This is what an STE is. It doesn't have to be out of this reality like an NDE. This is why many of my cancer patients told me they were now grateful to their disease because it taught/showed them what I am writing about here. They didn't have a tunnel experience or see the Light but they had all the aftereffects. Their take on reality became a new Reality with a capital R.

The near-death experience and other triggers for spiritually transformative experiences are an initiation, but they are only the beginning. Next usually comes a deeper search for our Real Self, our connection to others and God. I've come to realize in my 40 years of searching that the Universe is benevolent and meaningful. We have to heal ourselves enough to be able to realize that and live our lives according to what we learned. So the initiation becomes a journey and as the modern holy book *A Course in Miracles* says,

"The journey to God is merely the reawakening

of the knowledge of where you are always,

and what you are forever.

It is a journey without distance to a goal that has never

changed." - *A Course in Miracles*

Later, in chapter 12, I will say more on

Spiritually Transformative Experiences.

In their detailed article titled

The endocannabinoid system and the brain

Raphael Mechoulam PhD and Linda Parker PhD concluded:

"Cannabinoid research was originally initiated with the limited aim of understanding the action of an illicit drug. After the chemistry of the plant and the pharmacological and psychological actions of THC were elucidated—or actually only assumed to be elucidated—in the 1960s and early 1970s, research in the field waned. However, over a decade starting from the mid-1980s, two specific receptors and their ligands [i.e., chemical activators]—the bases of the endocannabinoid system [ECS]—were found to be involved in a wide spectrum of biological processes. This ECS has opened new vistas in the life sciences, particularly in aspects associated with the CNS [central nervous system].

"One of the main results of activation of the presynaptic CB1 receptor is inhibition of neurotransmitter release. By this mechanism the endocannabinoids reduce excitability of presynaptic neurons. CB1 receptors are responsible for the well-known marijuana effects as well as for effects on cognition, reward, and anxiety. In contrast, a major consequence of CB2 receptor activation is immunosuppression, which limits inflammation and associated tissue injury. Enhancement of CB2 receptor expression and/or of endocannabinoid levels has been noted in numerous diseases, including CNS-related ones. Thus, a main result of CB2 receptor activation seems to be a protective effect in a large number of physiological systems."
Annual Review of Psychology 64: 21-47 January 2013

3 My Experience as a Patient with Medicinal Marijuana

"Medical marijuana could replace addictive prescription painkillers such as Oxycontin." [6]
John Craven MD

According to a Thomas Jefferson University medical school study on cannabis use to reduce opiate withdrawal symptoms, cannabis use before and during treatment decreased the patients' score on the Clinical Opiate Withdrawal Scale (COWS). This is a scale used to objectively verify withdrawal symptoms in opiate dependent patients. The lower scores indicate that cannabis plays a role in reducing the symptoms of opiate withdrawal. [6] But it is CBD that helps here, not THC.

I personally have experienced this first as a patient and then later as a caregiver. I first suggested it for my patients who were wasting away from the aftereffects of chemotherapy and radiation, and then later to help them while they were being withdrawn from opiates that also have side effects of nausea, diarrhea, anxiety and pain.

I knew it was risky for me professionally. I did this because I had used it while in the body cast and knew it could help the people I cared for.

---

6   See also online e.g., ncbi.nlm.nih.gov/pubmed/23795873 and Appendix 4 for comparison of MM to Codeine for pain.

My weight loss had come from two years of pain pills and muscle relaxants to treat my back pain. Through all of this, I was anorexic. I had almost no appetite. I couldn't eat much because the pills were not only numbing my pain, they were numbing my natural feeling of hunger.

I was in the body cast for seven months. My life consisted of waving to my children as they stood in my bedroom doorway and spending my time in bed trying to read. The opiate pain pills made it almost impossible for me to read. Three months into the body cast I threw away all my pills and started smoking cannabis. Looking back on that moment, it was the beginning of my new life.

How MM helped me

Here at this junction of my recovery is the pivotal truth of why I see the benefits of using cannabis selectively and carefully. Opiates took away my physical pain and appetite. Benzodiazepines as "anti-anxiety" agents and muscle relaxants (brand names Valium, Xanax and more) made me so relaxed and numb that I had no need to try to do anything for myself. And as nearly all psychoactive drugs act and do, the opiates and benzos require higher and higher doses as my brain and body became tolerant to them. Tolerance when we are taking strong prescription drugs is the worst problem of all because we keep increasing the dose to get the same results that we used to get. I was almost "skin and bones" so that later when the body cast came off, I was wearing my 12-year -old son's blue jeans and he was thin.

Soon I realized that I actually didn't need pain pills.

What I needed was a way to cope with my sad and dysfunctional life. But I was so drugged and psychologically and spiritually numb that I couldn't figure that out. What I had wanted when I took opiates and benzos was a way to alter my state of consciousness so I could bear the emotional pain of being encased in plaster and all my resulting frustration. By using marijuana I had that altered state and none of the anorexic side effects of the drugs. My head actually felt progressively clearer and my mood became lighter.

I was experiencing something expansive from using the cannabis that I wasn't getting from any prescribed drug. No one called it "Medicinal Marijuana" back then. Botanists and scientists call it Cannabis. Later on I learned from healers called Shamans that seeing its spiritual side by referring to it as "Santa Maria" can intentionally take us into a different level of relationship with it and its properties.

Coming off the opiates and benzos created a new problem. With my head clearing, my feelings and emotions came tumbling up into my awareness. It was as though a dam broke and I became flooded with everything I hadn't felt since the beginning of my back injury two years earlier because I couldn't feel on opiates or benzos. The MM helped me to get some objectivity to my situation by letting me see my feelings as transient. I had an insight that "this too shall pass."

This timeless state that MM kept taking me back to were scenes in my near-death experience. I hadn't gone there too much before using it because the benzos and opiates that I had been taking had me so numbed and confused. Nurses and later a psychiatrist told me that all those experiences when I was lying in the circle bed were hallucinations. The psychiatrist told me I was depressed and gave me anti-depressants which I knew I didn't need. I was numb on these drugs and overwhelmed by my painful life. I was not depressed.

Having stopped the drugs, my mind was clearing and I was experiencing that the MM was acting as a muscle relaxant. Using MM, after a few hours of its helping me stay in the Now, and at the same time, lifting me out of my worries, I fell back, totally relaxed. Some people who are doing this for recreation complain that it drains them, but if using it is for medicinal purposes, that draining can be a gift. When I thought of and saw the marijuana as my muscle relaxant and focused on that, I could let go and relax.

Without the disabling drugs my head cleared and I started connecting with the sacred beauty of my near-death experience— with the help of MM. And my friends who loved me brought me this illegal plant already rolled in paper so all I needed to do was light it and inhale it. I never over did it. One inhale around 2 in the afternoon and sometimes one in the early evening held me until I drifted off to sleep later that night. I could now watch movies (and follow the plot – because I wasn't overdoing MM). I read books such as Ken Keyes' Handbook to Higher Consciousness, Abraham Maslow's The Farther Reaches of Human Nature and Alan Watts' This Is It on Zen and Spirituality. I also started reading books on Quantum Physics and I could start to understand where science is beginning to explain the possibility of how these experiences are happening. I also began to realize that I had a future, that my pain and confusion would be over and I could survive and possibly someday thrive. The timeless-ness that MM created felt very much like the timelessness I experienced in my NDE. It put me in that timeless floating state that made it easy to go back.

The first experience I remembered was a feeling of lifting out of my body and then going off into a tunnel where my Grandmother was waiting for me. When I felt relaxed from the MM I experienced in this reality what my grandmother's love felt like. I felt my grandmother hold me tightly like she did when I was a child. I felt in my heart a comfort about how she let me know she would always be waiting for me when I do die someday.

When I used MM I remembered when I was out of my body looking down at the pathetic me lying in the circle bed. I actually felt that Compassionate Intelligent Energy holding me and I heard myself saying, "No Wonder, No Wonder." I knew that meant "No wonder you are the way you are now. Look at all the pain you went through as a child."

I remembered and could actually feel again this benevolent Energy, God's Love, holding me tighter. This loving Energy, that some call "The Holy Spirit," let me see my life through ITs eyes and feel myself again through ITs heart.[7]

I am not demeaning God when I use these words.

Remember that until my NDE I had been a longtime atheist, and these words helped me make sense of what I was experiencing. There was a huge shift in me – not the me that was a withering body lying somewhere else —but the Real me—what I now call my eternal Real Self, my Soul. The shift felt like all my atoms and molecules were moving over one frame – electrons jumping orbits.

---

7   I refer to God in a number of ways throughout this book. No matter what I may call "It", It with a capital I, is always the same One God.

I felt awake and totally expansive.

This whole new way of being, even though I had been trapped in a 30 pound cast made my life turn around. The opiates and benzodiazepines had clouded my thinking and from that I felt hopeless and down on life and on myself. The MM connected me to light feelings of love and even helped me to forget that I was encased and almost physically frozen in plaster. With the help of MM I felt connected to the Now, the moment where our real life is happening.

Being in the "Now," I could sense the Unity of all things, not the separation I felt when I was on drugs. I realized that how I felt about MM revealed how I felt about life—did I want to see the world in its "wholeness" or divided in a state of separation. Did I want to feel the beauty all around me or wallow in pity for my situation? MM showed me this all as I lie there "naked" in my inner life. (See diagram of Inner Life in Chapter 6 on page 57). All this and my new learning and insights showed me what I believed about myself and how I could find and make new choices. It also showed me how I ran from my Real Self, my feelings, and even ridiculed myself. All this turned my inner life upside down and got me ready, once I was able, to take personal growth workshops, read books on self-actualization and finally to go back to college and continue my education. Looking back on it all now, I needed that time to get to know myself again because even though I was a good mother—the rest of my life was out of control. Somewhere in the materialism I had lost myself, from childhood into the present, I had given myself away to others.

After having the profound life review during my NDE, I was able to see both me and others in my life as being on a hamster wheel, running but not communicating. Something in me, some new part that I had lost early in my crazy childhood was back and it helped me to somehow know that everything was going to be all right. That I was on a "new road." And even though my "close others" could have never understood my new awareness and consciousness that I was getting to know – it was all right that they couldn't understand. It was safer for me to keep it to myself until I got my strength back.

About six months after the body cast came off, I was strong enough and began attending personal growth workshops. The first one was called "Changing," run by a psychologist for 8 women who like me were standing on the edge of an emotional cliff and not knowing what to do about it. That class gave me the courage to enroll at the local University. The Women's Center on campus was running personal growth programs that I enrolled in beside the regular classes to eventually earn a degree in the health care field.

When working in near-death research, I met and became friends with Dr. Raymond Moody, author of the first bestseller on NDEs called Life after Life. He likened my long months in the body cast to being like the cocoon with the chrysalis then coming out and becoming a butterfly. Had I continued on the opiates, benzos (and other assorted drugs I was given) I never would have had the many realizations that I needed to look at myself and my life and do something about it. My spinal surgery was telling me that my life was breaking my back. It was holding me down because I was holding myself down and not standing up for myself.

Medicinal Marijuana helped me to listen to my symptoms and to begin to sort them out. It helped me to allow myself to start to dream about the dreams I had given up when I got married and became a stay-at-home mom. Still, happy that I did that for myself and my three children, they were in school now. And I wanted to go back to school too. I realized that I wanted to pursue my dream of being a nurse or some kind of health care professional. MM let me see behind the curtain that was made of "Can not" and "Should not."

The Power of Pot

John Craven MD wrote:

"I have read that marijuana is only indicated for those who have 'not responded well to other forms of treatment.' This in my opinion is complete nonsense – driven entirely by misperceptions and prejudices resulting from its status as an illegal substance used outside the mainstream of medical practice—and denial as to the true risks of many other drugs commonly prescribed today." [8]

In his 1971 landmark study [7] of marijuana users, the pioneer consciousness researcher, teacher and author Charles Tart PhD summarized that cannabis use causes a reorganization of our mental functioning. Just how and to what degree this effect and experience may be depends on five basic factors: a particular person taking a particular drug in a particular way [i.e. delivery method] under particular conditions at a particular time. This is another way of describing that the experience of using cannabis is based on set and setting, whether one is using it as MM or recreationally, socially or casually.

---

8   Craven J 2014 *The Power of Pot*: How to Harness the Medicinal Properties of Medicinal Marijuana (Kindle only available at Amazon.com)
9   Tart C 1971 *On Being Stoned*: A psychological study of marijuana intoxication. Science and Behavior Books.
Online free at druglibrary.org/special/tart/tartcont.htm

Person + Drug + Way + Conditions + Time
= Set and Setting

Cannabis with enough THC increases most of our sensory experiences and in a low dose can help us stay in the present moment, often called "The Now." Using cannabidiol (CBD) alone that has very low (0.3%) or no THC is calming and "Now-orienting" (see chapter 11 and Appendix 6 and 13 for details). Used in a positive setting MM can be more effective than other drugs to help critically ill people to be able to say what they want to say to their significant others. If they are preparing to die and are already spiritual, they may want to be awake and alert during the time they have left.

Waking Up

Since my near-death experience 40 years ago, I have searched for the truth about who I was in my glimpse of eternity. (See Chapter 12, Map of the Sacred Person).

I now know it is the core of who we are and who we will continue to be after our body dies. This part of us that I feel comfortable calling my "Soul," knows and communicates to my heart that there is a capital S Spirit Higher Power/ God-Goddess-All-That-Is operating in the Universe which gives meaning and purpose to our life.

That Spirit in Its wisdom gave us the plant kingdom to enjoy, commune with, and help us. And when we take some plants into our body, our energy takes in their energy too. How we take them in, our attitudes, our deep held beliefs, and most important, our emotional condition at the time of taking them in, has a great deal to do with giving them every chance to help us. If you the reader are new to this, you may not believe some of what I wrote above. And what you are about to read. And that is all right. All we need do is have the intention that we are willing and open to whatever will help us.

As I stopped using the prescription drugs and started smoking MM, I immediately started to transition from a drugged out "turtle" of sorts to a new level of "hyper timelessness" where things became more interesting. I wasn't smoking the super strong buds that have emerged over the last 20 years. By contrast, I was given the leaves that contained enough effect to help me open to a gentle new peace. And I needed only one inhalation (usually) which lasted me all day. I had the "set and setting" for one dose, which I will go into in a later chapter.

During this time in the body cast, I had my first experience writing poetry. While using MM I felt a similar timelessness that reminded me of what I experienced in my NDE.

Timeless Zone

I'm sitting in this reality

Somewhat confused

I have resided in more than one dimension And physical reality puzzles me more Than the "altered states" I have played in.

Confusions arise here

As beliefs drop away.

All the dichotomies come crashing down

And opposites dance on the same continuum.

Sometimes, not to be recommended

While going through pain.

But probably the quickest way to go.

To simply exist is not to live

But just "to exist."

To live, really LIVE, We have to dive in And swim around in IT!

When the Doors of perception

Finally clear,

The awareness of living in the NOW.

Timeless NOW.

So this is "Cosmic Consciousness."

First it comes in flashes

Then interludes.

Finally, an understanding of GRACE.

And what GOD is—

Where I wind up at the end of all my words.

BH

Carl Sagan on Its Prohibition and Illegality

"The illegality of cannabis is outrageous, an impediment to full utilization of a drug which helps produce the serenity and insight, sensitivity and fellowship so desperately needed in this increasingly mad and dangerous world."

Carl Sagan PhD astronomy and astrophysics at Cornell, Harvard, and the Smithsonian Astrophysical Observatory

Comment: From the likes of Harry Anslinger in the 1930s to Bill Clinton's appointed drug czar to the DEA and NIDA's cannabis drug policy still active today, this illegal law Sagan points out has *prevented* scientific and medical *research* on this potentially helpful natural plant 'drug' for over 80 years. It still prevents scientists and clinicians from helping us learn more about its healing properties.

We can each help by voting for pro MM laws in our states.

Sagan quote is from Holland J ed. *The Pot Book:* A Complete Guide to Cannabis. Park Street Press, Rochester, Vermont page 395. See also azarius.net/news/306/ and Carl_Sagans_essay_on_cannabis/http://marijuana-uses.com/mr-x/ or an online Startpage.com search for his quotes for more by Carl Sagan.
He died in 1996 at the age of 62 0f pneumonia and a bone marrow illness.

4 More Secrets of Little Plants and Big Trees

What Science is Proving

D r. Suzanne Simard at the University of British Columbia is concluding that "trees are interacting with one another in a symbiotic relationship that helps the trees to survive. Connected by fungi, the underground root systems of plants and trees are transferring carbon and nitrogen back and forth between each other in a network of subtle communication. Similar to the network of neurons and axons in the human brain, the network of fungi, roots, soil and micro-organisms beneath the larger 'mother trees' gives the forest its own consciousness."

In this four minute video [10] it explains how the forest has its own consciousness. It is amazing and of course, hard to believe, when I was an atheist. However, after my near-death experience I was experiencing an awareness of a Higher Power/God/Spirit operating in the Universe which was giving meaning and purpose to all existence.

---

10   4 minute video on this research by Dr. Simard expandedconsciousness. com/2014/03/07/the-consciousness-of-trees- video/

This God/Spirit/HigherPower is already and always everywhere present at all times. And now as science is proving, It in the forests and in all plant life.

The above information may sound suspicious at first but fungi actually have a sophisticated biology of their own. For example, the neuroscientist Sam Wang PhD of Princeton University describes how fungi called "bread mold" (Rhizopus Stolonifer) even in the absence of light have a circadian rhythm of their own and are entrained [linked, joined, drawn along] and influenced by the light.

Healing in Our Times

In 1981, I attended a conference in Washington DC called "Healing in our Times." The brochure for this particular conference peaked my curiosity. The speakers included 1) Elisabeth Kubler-Ross MD who had just written *On Death and Dying*; 2) Elmer and Alice Green who were investigating how Biofeedback worked and created a group to study subtle energies;
3) Robert Becker, an orthopedic surgeon who from his research discovered the natural bio-electric energy that aided in the healing of broken bones and wrote the book *The Body Electric*: Electromagnetism and the Foundation of Life; and
4) Delores Krieger RN, PhD who created the skill of Therapeutic Touch in a Master's Degree program for nurses.[11] I took two workshops from Dr. Krieger.

I was already noticing that something positive was happening to my patients, and to me, when I touched them.

---

11   See my first book for a more complete description of this conference. *Full Circle*: The Near-Death Experience and Beyond.
Simon and Schuster, Pocket. New York, NY 1990

I noticed a transfer of energy that I didn't understand until I attended her workshops at this conference. I was also fascinated by 5) Dr. Bernard Grad's presentation, about researching plants and their ability to be able to pick up the feelings of the humans around them. I met him in the elevator of the hotel and told him that as a child my grandmother taught me what he had explained in his presentation about communing with and the sensitivity of plants. He chuckled and told me that I had a wise grandmother." Bear with me while I talk about her. [12]

My Childhood

When I was a child from 4 years old, I often stayed at my grandmother's, and she sometimes taught me about her plants. Her home was a small simple one in a busy neighborhood of working class Detroit. The Boy's Club was across the street and I could hear streetcars clanging nearby as I sat on the front porch. On the railing of her little wooden porch, neatly lined up were several kinds of plants. Some were flowering and some weren't but she loved them all the same. In her back room that served as a root cellar and sewing room, she had more plants, and some cuttings that were "drying" and she taught me to love them too. She warned me never to water them if I was angry or sad. "They will know and it will hurt them." Sometimes she sang as she watered, pruned and admired her perennials that she moved into the back room in the winter. Detroit's winters are

---

12   In his presentation on his research at McGill University in Canada, Dr. Grad explained to us that he looked at the growth rate of seeds and plants which had been watered with water held and treated with energy by different people (including the healer Oskar Estebany, a lab technician and a depressed patient) and compared it with the growth rate of seeds and plants where untreated water was used. He found a significant increase in the growth rate of seeds and plants exposed to the "treated water". He found that when healing energy was involved, there was an identifiable difference in positive outcome greater than the control in which no healing energy was used. This research and others were done from 1957 to the late 1990's.

usually overcast, with little sunshine, but all my grandmother's plants held on and even flourished.

She taught me how to commune with these living plants.

I didn't know at the time that I was meditating with them, but I was. I slowed down to where the plants were. I found this to be peaceful and lasting. I was being in the Now of my "awakened awareness" and I found joy there. Later, I found books that confirmed what I was experiencing. It may seem hard to believe, but I learned that plants are conscious beings. They may not be in the same level of consciousness as we are—but more recent research is validating little plants and big trees as being conscious.

Here are some recent Scientific findings in plant research, as summarized by Jon Lieff MD on his blog:

"Taken together, these many different capabilities and behaviors show that plants do, indeed, have intelligence. They plan for the future, make decisions, move, evaluate, send complex signals and work cooperatively with other plants and with microbes. They have a wide range of senses and a wide range of behaviors, including manufacture of many rare chemicals for special situations.

"Because plants operate at a very different speed and with very different behaviors than we are used to, it is very difficult for humans to understand them."

A previous post in Dr. Lieff's blog discussed the fact that, "Humans can't understand the point of view of animals who have different types of brains, with different senses and motor capacities. It is much less possible for humans to understand the life of the intelligent plant." [13]

Vibration as Another Way to Communicate

Six years after my NDE (1981) and a year after my respiratory therapy (RT) training, I was working in critical care units as a respiratory therapist. At times I noticed subtle energies in my patients that we weren't taught about in RT school. For example, I had experiences with dying patients and I felt comfortable helping them adjust to the process of leaving their earthly life. In many, I felt in a subtle way their energy leave their bodies as they transitioned. Actually it was becoming apparent to me that in the last few days of their lifetime, their physical energy weakened and transformed into spiritual energy that could affect or at times "entrain" their loved ones and caregivers. (To entrain is a subtle process by which one event influences or effects another event or process.) An example would be tuning forks. When one is struck, all nearby tuning forks will identically vibrate.

So the one vibrating tuning fork entrains the other tuning forks. I had no clear answers for this phenomenon (that I have written about in my previous books.) But science is getting closer to acknowledging that entrainment exists.

As a foundation, Dr. Lieff writes on his website:

"Everything in nature vibrates and produces sounds - from sound waves of vibrating molecules to black holes.

---

13   See more at: jonlieffmd.com/blog/plant-intelligence- primer-and-update#st-hash.PbFd1Fxh.dpuf

Sound vibrations are everywhere." [14]

We may not be able to hear the vibrations but they do exist. And like the tuning forks, the strongest vibration entrains the others. So perhaps when we are out in nature, we are being entrained into the vibrations of the trees and plants. We also are inhaling the clean sweet oxygen (and negative ions) that the plant kingdom is gifting us.

Growing Medicinal Marijuana

So now that we know more about the research into the secret life of the plant kingdom, if you qualify for a MM license, growing your own Medicinal Marijuana may be a pleasurable endeavor and even joyful process.

I received the following email recently from a woman in Florida who heard I was writing this book:

"When it becomes legal in our state, I'm going to get a MM license and grow my own. It's not just fun for me. I know this will be an amazingly different kind of experience raising a plant that is going to help me.

"I fell a couple of years ago, hard on both knees. I suffered for eight months with bone bruising (edema or fluid in bone marrow of both patellae).

And that left me with arthritis in both knees. I was given cortisone injections, Supartz injections, opiates, muscle relaxants and

---

14   Jon Lieff MD in Unique Effects of Music on the Brain Searching for the Mind Blog jonlieffmd.com/blog

anti-inflammatory drugs. Knowing what I do about these

"hard drugs" (opiates) I used them sparingly.

She concluded: "When I get to grow my own MM plants, they will be in our home. I will feel again what I felt growing up on a farm with the summers filled with wonderful fields of growing food."

Willie Nelson, the singer, said "Marijuana is an herb and a flower. God put it here. If He put it here and He wants it to grow, what gives the government the right to say that God is wrong?"

Details on growing MM is beyond the scope of this book.

However there is a variety and amount of information available online that comes from sincere, non-marketing sources and also those who are in business selling information and growing products.
See, e.g., calgarycmmc.com/growingtips.htm#931395472

Excellent DVDs and YouTube videos online are also available to show us countless details.

I offer a few additional principles. In the best of circumstances successful growing can take a few months. But each farmer can, through trial and error, determine their own pace. A basic principle is first having a written plan, then executing that over time and having patience for your plants to grow to maturity.

A beauty of raising these plants ourselves is the experience of knowing the plant at a more personal and deeper level. If we the grower wants MM or reframes it as Santa Maria, we can intensify the experience by praying over the seeds when we plant and care for them over the long weeks and months. Before each watering we can hold the watering can with our hands on each side for a few minutes so that our loving and positive energy infuses the water of life and growth that they need.

Ask yourself what you want your MM to be and say that in your prayers. During the growth period again use your hands on both sides of the plant(s) and say a prayer.

Include intention, positivity, gratitude and love.

This is the same attitude that we can choose when we use MM in any of our healing process for our self or our patients.

If you are growing plants for a loved one—visualize them smiling and healthy and at peace. Intention and visualization are powerful tools to energize the plants. When harvesting, do it when you are calm and hopeful for your plants' abilities. Throughout the growing and during the harvest play soft calm music. You may find that through all this process you experience an intimacy with your plants that touch you in deeper ways. We love growing these plants because they are going to be helping us and the people we love.

What I have heard from caregivers who grow a few is that the plants in return show them the oneness and balance when we commune with nature. They say that raising the plants is a healing experience for them and they will always look at all plants and trees with a new respect and love. They may also admit that they needed something like this to help them with their own level of stress and that raising these plants was a joy.

Part II The Clinical Side
or How to Use It

Just as an acorn contains the potential for its identity, a giant oak tree, we, too, contain the potential at birth for our soul to develop to its fullest in this lifetime.

–Barbara Harris

As shown in the Timeline in this book's rich Appendix, there are some 20 States where Medicinal Marijuana is legal, 4 plus DC where use is legal with an MM provision, 4 where possession is recently decriminalized (Nebraska, Ohio, N Carolina, and Mississippi). All this happened in about 19 years since California's compassionate MM act in 1996. This leaves only 20 states to enact one or more of the above to free us from its being illegal.

London's *The Economist* wrote "Americans spend an estimated $40 billion getting high each year, about 20% of what they spend on cigarettes and alcohol. The legal market, estimated at $2.5 billion last year, is just a fraction of that but will grow. A majority of Americans (52%) are now in favor of legalisation—in 1969 that figure was just 12%. California is likely to legalize cannabis in 2016; weed-watchers reckon that federal legalisation may be five to ten years away."

Legal status
As of January 20th 2015

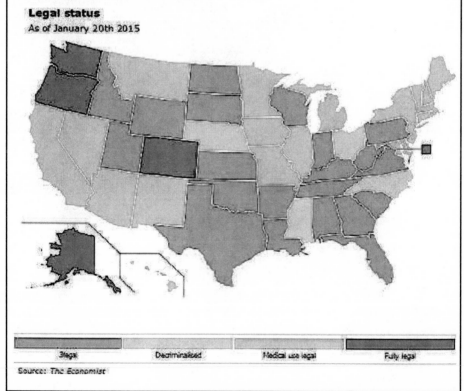

| Illegal | Decriminalised | Medical use legal | Fully legal |

Source: *The Economist*

5 Benefits of MM

"Sometimes just after one puff, patients' pain scores will drop by something like 70 to 80%."

Sunil Aggarwal MD, PhD [15] member,
New York Academy of Medicine

My first position in health care was in a hospital as a respiratory therapist in 1979. I was 36 and then went on to do home care for the dying. During this time I was published in nursing and respiratory care journals on a subject I coined as "The Emotional Needs of Critical Care Patients." I was asked to teach on this subject at Nursing conferences and hospitals. Finally, I became a full time psychiatric researcher in the medical school at the University of Connecticut, a position I held for six years. I got to see and know how the system went.

The past medical model as I have understood and observed it, tends to elevate the opinion of the physician (the expert) automatically putting the patient in a passive role. The patient is forced into "behaving" so they can fulfill expectations.

---

15   The Future Of Marijuana In Clinical Practice: Q&A with Dr. Sunil Aggarwal
leafscience.com/2013/09/24/inter-view-dr-sunil-aggarwal- future-marijuana-clini-cal-practice/

I sadly observed psychiatry stop doing and teaching psychotherapy and instead suppressing symptoms with drugs and sometimes ECT (electro convulsive [shock] treatment). In concert with the pharmaceutical industry there has been a profitable alliance between physicians and drug companies that has blurred all boundaries. "Big Pharma" has infiltrated scholarly medical journals, medical conventions, and medical schools. The results continue to suppress any investigation of both 1) what symptoms mean and 2) how to help the patient find a new *balance*, or *homeostasis*, which "Holistic Health" calls for in body, mind and spirit. Unfortunately, I have also observed many physicians accept this paradigm of suppressing symptoms with drugs as their norm and not trying to go deeper. They find a suitable name (a "diagnosis" or more accurately a label) for people's problems, try to suppress their symptoms with drugs and believe their job is finished.

Reviewing what I just wrote above, I am still writing about the emotional needs of patients here in this book.

Taking back our Power

By contrast, from working in both the recovery movement and holistic health for 23 years, and now watching the introduction of Medicinal Marijuana – I have seen that within these healing approaches the patient takes responsibility and becomes much more active in their own healing. By using MM the patients have more choices for how to best work with it, a major choice being to take charge of using it selectively and carefully for their healing. As you read through this and other books and literature notice how you can do that.

You may also notice that other prescription drugs for chronic pain are not always necessary with MM available. At other times the opiate drugs' dose can be tapered, which CBD can clearly help taper. And it helps for tapering other drugs.

We have more choices to lessen our pain. A positive choice is that we not deny, repress, drug or ignore our symptoms. Others and I have come to understand that our symptoms are trying to tell us something, as I discovered in my long experience when I was trapped in the body cast. I listened to the various parts of my life. While I was trying to make sense of my predicament I learned that underneath my spinal fractures, magnifying my pain, and not being the only source of it, was my experience of not having lived authentically from and as my Real Self for so long.

Listening to our Symptoms

"I like the fact that in this book you're recommending marijuana not as just an anti-anxiety medication, but as a vehicle to help explore the underlying issues."

Bruce Greyson, MD [16]

Psychoneuroendocrinology (PNE) is a field that studies how emotions and behavior interact with hormones and the nervous system. [17] A related discipline adds immunology as psychoneuroimmunology (PNI) to address the interaction among psychological processes and the nervous and immune systems.

When we combine these with what we have learned from some 120 years of experience in psychodynamic and trauma psychology we can begin to understand how finding and addressing unresolved issues can help us to a deeper healing (see Stages of Recovery

16   Personal Communication, Bruce Greyson, MD Emeritus Professor of
        Psychiatry, University of Virginia School of Medicine
17   reachingbeyondnow.com/2010/05/26/what-is-psycho-neuroendocrinology/

in the next chapter). This understanding shows that underneath common chronic illnesses (see 'Stage Zero illness' next chapter), there are trauma effects or wounding that may not be causing the dis-ease or disorder directly, but have weakened our body systems so that wherever we are most vulnerable becomes "out of balance" or "diseased" and makes us potentially "sick." In other words, when we are often stressed and some of this is coming from unresolved issues with buried emotions, this wears down our immune and other bodily functions. The weakest link in our organs starts to break down. By treating this weak organ or system, we may be able to resolve or at least medicate the chronic problem. However, there might continue to be one health problem after another. We need to go deeper to find the underlying stressors that weaken our body.

Points to take home: What I am saying throughout this book has to do with the wounding that underlies our too often clouded consciousness that thereby keeps our body, mind and spirit out of balance. For some, using Medicinal Marijuana can help us unlock important parts of this unfinished business. This may seem painful at first. But in the long run it becomes an attitude of "I want to find out more about what happened and what's inside me." This is because there is a substantial emotional, physical and psychological release each time we have a realization that some call an insight and when one is in the extreme an "epiphany." The Adverse Childhood Experiences (ACE) study and countless similar detailed clinical scientific reports show that most of us have experienced several to many hurts, losses and traumas that have affected our ability to navigate our life as smoothly as we would like.

There are some lucky people who are free of earlier wounding or traumas. In the absence of repeated serious physical and/ or emotional trauma, some of us seldom get sick with bothersome or disabling chronic illness.

Medicinal marijuana can act as a kind of interpreter to help us understand what in our life may need to be examined, moved into balance and then healed. Healing comes from naming and working through our unfinished business. We can consider starting by using MM to explore our inner life. What could my symptoms be about? And how might I heal around them? Can I accept what has happened over my lifetime and consider letting any of it talk to me?

Using MM, to start, I imagine what might have caused this symptom or these symptoms. Then I imagine what nurturing changes I may be able to make in my life to facilitate my healing.

Some long-term MM users may already know some of this information. The war on drugs hid using this gift of expanding our consciousness to include our helping balance our health and expanding our awareness by possibly stimulating openings of our unconscious mind.

This personal inward examining can come through an expansion of our **Conscious Awareness** (which is Secret number **3**), then # **6**) What are my **Symptoms** trying to **tell** me?, and **8**) **Take charge**, and **become proactive** as introduced in Chapter 1.

If you may have a lingering image of two eggs in a fry pan and a voice saying "This is your brain on drugs!" don't let it confuse you.

Your brain will not come out sunny side up from using MM as I describe it in this book (but your disposition on life may). This "Brain on Drugs" was a ploy from many in our fearful and unknowing government. When used carefully as others and I describe as MM, this fear is not true. For at least 1500 years before the prohibition by our and other governments, cannabis was used all over the world by many as a healing aid. Ancient texts from long ago speak of the plant's gifts to humankind.

MM if used in the mindset I am explaining, will help you recognize what is helping you and what is not. It will help you go deep inside if you so choose and from a holistic health standpoint show you what is out of balance and can be changed. Or you also have the choice of just letting MM help you laugh, munch lightly and see the funny side of life that was hard for you to experience until now. It can take you out of the misery of your problems and lift your spirits. At the very least, it can help you to be distracted from your misery.

John Craven MD said, "I believe that time will tell that medicinal marijuana will result in far less risk than many other drugs commonly prescribed today. If anyone can benefit from the judicious use of this naturally occurring substance without resorting to many of the more powerful psychoactive drugs on the market today, both they and their brain will be better for it." [18]

I will again use my own experience here to illustrate what Craven said. When I spent seven months in a full body cast after a spinal fusion operation, I took myself off the heavy opiates and benzodiazepines, and with the help of loving friends, replaced those disabling and toxic drugs with marijuana that I experientially found was healing for me as I describe.

---

18   Craven J 2014 *The Power in Pot*: How to Harness the Medicinal Properties of Marijuana in the Management of Clinical and Stress Related Disorders. supportnetstudios.com. Kindle Edition

I was able to use MM to withdraw from the drugs. I now know that it was the CBD in what I smoked that helped. CBD in tincture or oil sublingually or oil vaped works to help taper or detox from these and other addicting drugs.

Addiction to benzodiazepines and opiates remains epidemic in this and other countries worldwide. And getting off of them is painful and takes a long time. As CBD, Medicinal Marijuana may help us get off most addicting drugs. If we may notice any withdrawal signs from using MM itself, it is easier to withdraw from (or use occasionally), when the patient is ready, which I discuss in later chapters.

MM has had many names over the millennia: pot, grass, weed, Mary Jane, ganja and the name botany and biological science gave it – cannabis. Modern shamans have called it "Santa Maria" because of the spiritual properties that emerge from it when we use our prayers and healing intention.

MM may have more healing applications than any other plant on this planet. But it is not in itself a cure. It is another tool to work with. Responsible use of it by engaging a mind-body relationship, we can open the door to a new and powerful healing that can take our journey to new depths and not just try to simply lessen our symptoms.

We can find a new understanding, a new freedom, a release of guilt or shame (Did I deserve this illness?). We may even find a tender intimacy that wraps us in a blanket of compassion for ourselves and where we are now. In this new relationship, the old stressors melt and we can use all of our energy for even more understanding of what our life may be about and how we can balance our body, mind and Spirit.

Worry-free Zone Meditation

This meditation is actually a visualization that if practiced every day can help us to recognize when we are worrying and letting whatever is troubling us go. Remember, worry does not prevent bad things from happening. All worry does is take us out of the moment or the Now where we want to be. Worry is the ego's way of strengthening its hold on us. Being in the Now is our True Self or Soul being in and connecting with this moment. [19]

Sit in a chair or lie down comfortably. Ask MM/Santa Maria to help you with this meditation. Visualize yourself writing on several pieces of paper about all your worries for today. You can write each worry on a separate piece of paper. (If you wish, you can actually write this out, not just visualize it.)

Visualize a shopping bag on a hook in your closet. Take the pieces of paper with all your worries on them and place them in the bag. Now, sitting or lying down, enjoy this moment of being worry-free.

Through the day you may "catch" yourself worrying. If so, visualize the bag in your closet and visualize yourself dropping a slip of paper with the worry in the bag. Visualize and see yourself walking out of the closet, worry free.

---

19   Whitfield CL 2013. *Wisdom to Know the Difference*: Core Issues in Relationships, Recovery and Living. Muse House Press, Atlanta, GA

Barbara Harris RT

Institute of Medicine Report from 1999

This was a remarkable study and report for several reasons, including because it came from a group of otherwise conventional academic physicians and selected government staff assistants. Under Bill Clinton who said he 'never inhaled' and his strongly anti-cannabis 'Drug Czar' Barry McCaffrey's watch the Office of National Drug Control Policy perhaps unexpectedly funded this study in response to public pressure to allow the medical use of marijuana. It examined many issues, including the abuse potential, tolerance, withdrawal, and gateway risks of medical use of cannabis. The remarkable outcome which the authors expected to be negative, surprisingly was positive. They could find nothing negative or wrong in the study about carefully using cannabis as a medication. Among some surprising potentially positive recommendations were:

"RECOMMENDATION: Short-term use of smoked marijuana (less than six months) for patients with debilitating symptoms (such as intractable pain or vomiting) must meet the following conditions:

• failure of all approved medications to provide relief has been documented,

• the symptoms can reasonably be expected to be relieved by rapid-onset cannabinoid drugs,

• such treatment is administered under medical supervision in a manner that allows for assessment of treatment effectiveness, and

• involves an oversight strategy comparable to an institutional review board process that could provide guidance within 24 hours of a submission by a physician to provide marijuana to a patient for a specified use." Mar. 1999 page 179

Comment - Enough people already know about this use so that many may bypass conventional medicine and go right to personal MM use. I still recommend contacting your physician if your medical problem does not improve rapidly and stay fully improved.

Joy JE et al 1999 *Marijuana and Medicine*: Assessing the Science Base, Institute of Medicine, March 17, 15 845-1132 See page 62 for more.

6 Two New Paradigms for Healing

Holistic Health

From 1991 to 1996 I had the privilege of working alongside Alan Gaby MD, a pioneer in the field of nutritional and holistic medicine. While working with him he was the president of the American Holistic Medical Association. My individual practice worked in harmony with his and our blend helped many to come into a natural balance. He oversaw the nutritional needs of the patient and I worked with them to learn how to breathe in a healthy way while moving blocked energy. The only way massage and energy balancing works is if the patient is breathing deeply, slowly and hopefully, in concert with the therapist. Many times, my patients would return telling me of dreams they had after the last session that told them either symbolically or blatantly what they needed to do to heal.

Holistic health is a new paradigm in healing that has gained ground recently in medical schools. The University of Arizona Medical School has an academic division investigating and teaching about alternative methods for healing with Andrew Weil MD as its Chief.

Other universities are beginning to look at this area too.

A primary philosophy in "functional", "alternative" and "integrative" medicine is balance. Dis-ease (absence of ease or health) is caused by an imbalance of body, mind and spirit. This paradigm includes, when appropriate, using aids as meditation, acupuncture, yoga, therapeutic massage, herbs, biofeedback and breath and energy work—in addition to selected allopathic (i.e., conventional medical) treatments. Patients can still work with the methods used in the old medical model but their physicians are encouraging them to try adding other alternatives.

If the patient connects with past trauma and/or neglect (see Stage Two Recovery below) there is the use of Emotional Freedom Technique (EFT) [20] and Eye Movement Desensitization and Reprocessing (EMDR) to help alleviate the stress behind the past wounding. There are several other treatments too long to list here. Some may work as a placebo effect but are totally acceptable if they help.

The Recovery Movement

From the 1980s through the 1990s an increasing number of people began awakening to their traumatic experiences and began to heal by addressing them. This phenomenon, called the recovery movement, is part of a new paradigm, which gives a new and expanded understanding and belief about the recurring

20   The Emotional Freedom Technique, or EFT, is the psychological acupressure technique I routinely use in my practice and most highly recommend to optimize your emotional health. Although it is still often overlooked, emotional health is absolutely essential to your physical health and healing - no matter how devoted you are to the proper diet and lifestyle, you will not achieve your body's ideal healing and preventive powers if emotional barriers stand in your way.
EFT is easy to learn, and will help you: 1) Remove Negative Emotions, 2) Reduce Food Cravings, 3) Reduce or Eliminate Pain, and 4) Implement Positive Goals.
see Mercola.com and search on startpage.com

painful part of the human condition and how to heal it. This approach is so effective and has developed so much momentum for so many people for two reasons: It is grass roots, that is, its energy comes from the recovering people themselves, and it employs the most accurate and healing of all our accumulated knowledge about the human condition. What is different about this knowledge is that it has now simplified and demystified the healing process.

A basic principle in the recovery movement regarding trauma effects and healing around them is to address it in the sequence of stages, which I describe below in Table 1 and following.

Table 1 Recovery and Duration According to Stages*

Recovery Stage	Condition	Focus of Recovery	Approximate Duration
3	Human/spiritual	Spirituality	Ongoing
2	Past trauma effects	Trauma-specific recovery program	3–5+ years
1	Stage 0 disorder	Basic illness, full recovery program	Months to 3 years
0	Active illness Underlying and contributing is unconscious wounding that is generating pain and/or conflict	Usually none	Indefinite

* Read this table from the bottom up. From Charles Whitfield 2005

STAGES OF RECOVERY

Understanding the stages of recovery is important in helping us heal. These are summarized in the above table and further explained below.

Stage Zero

Stage Zero is manifested by the presence of an active illness that is physical or a disorder such as an addiction, compulsion or another disorder. This active illness may be acute, recurring or chronic. Without recovery it may continue indefinitely.

At Stage Zero, recovery has not yet started.

Underlying the Stage Zero illness, weakening the physical body and sometimes playing "havoc" emotionally can be an unconscious wounding that will continue to cause problems if not addressed. This is where Stage Two recovery can not only help us to recover from an active illness or disorder, but help us to prevent more health problems and have a healthier and longer future.

Stage One

Stage One is when recovery begins.

It involves participating in a full recovery program to assist in healing the Stage Zero illness, problem, disorder, condition or conditions. In both Stages Zero and One— underlying the illness, a major causal factor may be past trauma. If the trauma was or is repetitive there may be Post Traumatic Stress Disorder that needs to be addressed. Just knowing about it will help the person understand what happened to them.

Some past trauma may not appear to have been extreme, but if it is an underlying factor in an illness it still could help to address and release the buried emotion behind it.

I found out while addressing my own past trauma that I experienced two different but related traumas each time. The first trauma was physical abuse and the second trauma was when no one would let me talk or cry about it. And that happened over and over again. Not being able to release our grief over being abused leads to a deeper traumatization that is stored in our body. This blocks the natural flow of our energy. I call it the "clenching" disease (which MM may help release). We are so tight that the circulation of blood and lymph in our body has a difficult time circulating. That is why this new paradigm embraces massage, meditation and yoga plus other alternative aids that help to move our bio-energy and lymph flow.

Stage Two

Most people may try to bypass Stage Two. It involves healing the effects of past traumas, sometimes called adult child or co-dependence issues. In this context, an adult child is anyone who grew up in an unhealthy, troubled or dysfunctional family. Many such adult children of trauma may still be in a similar unhealthy environment, whether at home, in one or more relationships, or at work. Once a person has a stable and solid Stage One recovery—one that has lasted for at least a year or longer—it may be time to consider looking into these issues.

Stage Two is where my husband and I as co-therapists

work closely with patients doing individual and group psychotherapy. This is rewarding work for both group leaders and group members. Some members stay a long and appropriate time. The group becomes a safe family that helps us to understand and experience what healthy relationships are. We ask for a one year commitment when the member starts and have observed that 3 to 5 years is an ideal time to use group therapy to help heal. For over two decades, we taught this new paradigm I am describing at Rutgers University's Institute for Alcohol and Drug Studies to therapists of varying degrees including alcohol and drug counselors, social workers, psychotherapists, psychologists, psychiatrists and clergy.

If we try to bypass this stage, (to address these previously unknown issues from our past), if we try to ignore the lower to get to the higher levels of our consciousness, something— we can call it our ego (baggage, unfinished business), shadow (Jung), or repetition compulsion (Freud)—will pull us back until we work through our particular unfinished business. So Stage Two is about listening to our inner life over time and telling our story to safe people.

Our Inner Life

In 21st century Western society we are constantly busy.

We work. We play. We consider "just being" as being lazy. We have computers, phones, tablets and more to constantly fill our attention. Using Medicinal Marijuana is the opposite. It gives us the opportunity to slow down and listen—to the silence, or if we choose, to our own inner life.

Asking a patient about their important inner life while they are using Medicinal Marijuana can open up a new level of intimacy for them with their self, others and God. Taking deep relaxing breaths and going inside to listen to our gentle (or loud) voice within has benefit beyond texting or emailing everyone else. We get to hear and become familiar with our own thoughts, feelings, wants, needs and all listed in the diagram below of our inner life.

Figure 2 My Inner Life

-Beliefs   -Needs   -Thoughts
-Feelings   -Wants   -Decisions

-Choices   -Experiences
-Dreams -Creativity   -Fantasies
-Sensations   -Intuitions

-Higher Self   -Primal feelings
-Unconscious experiences
-Insights and revelations

-Higher sense perceptions
-Unfinished business
-Connection to God

My Inner Life - It is important to become fully aware of what is coming up for us from moment to moment in our inner life, to observe our own heart and mind at work.

For many people this will be the first time they realize they can listen and learn more about who they really are. I have heard, "This is a whole new world for me—just to sit quietly and hear my own needs, especially my fears and regrets (aka, guilt and shame)." I explain that the painful feelings that they are hearing don't need to stay in them. And if they like, we can talk about each one and figure out a way to let them go.

When they get comfortable with this type of discussion I introduce the idea that when we listen to ourselves, this is the beginning of expanding our consciousness because we are becoming aware or conscious of who we are, what we want to keep about ourselves and what we want to change.

Inner Life Well-Being

When someone is diagnosed with cancer or other life- threatening illness, the focus needs to be not just on their physical state, but on their emotional state as well. Perhaps even more so. Ideally, they may benefit by asking these questions:

•Do you love and value yourself?

•Are you at peace? ...happy?

•Are there people in your life who matter to you and to whom you matter?

•Do you feel your life has purpose?

•What are you passionate about?

•What brings you joy? [21]

Each question could be the start of a personal session.

Or each question could be a creative writing experience where there is no pre-planning but a steady flow of whatever is coming up, while using a low dose of MM or not using any.

Stage Three

Stage Three is the experience of natural spirituality and its incorporation into our daily life by feeling a deepening connection to our Higher Self and Higher Power. (See Map of the Sacred Person on Page 130 in the chapter on Spiritually Transformative Experiences.) We make meaning out of our past hurts, losses and traumas. They no longer contain an emotional charge and become part of our story that we don't need to worry about anymore.

This process is ongoing. An example from my past is that I don't need to go over and over the abuse and neglect I received as a child. But because of the abuse and neglect I have become a therapist aiding others who also were abused. I have written books on healing from abuse that have helped others. If I hadn't been abused, I would likely not be doing the compassionate work I am doing now. I know the journey my patients are on because I have traveled that long and painful path too. I found that what was an early curse in my life has evolved into a gift, an accurate and constructive reframe for me.

---

21   Excerpt from Breast cancer article by Anita Moorjani Huffington Post 23 Oct, 2014

Sasha Shulgin PhD

Cannabis has been said to have 'quasi-psychedelic' effects when used in higher doses in some people, depending on their set and setting. In 1986 I met this pioneer biochemist at a small three day conference addressing the therapeutic and healing uses of psychedelic drugs. He rediscovered MDMA and introduced it to the therapeutic community.

Regarding how to use cannabis, he told me [22]

"You hold the marijuana in the palm of your hand looking at it. You ask it, 'Can you be this for me?' Whatever this 'this' is, let the patient decide what he wants from this sacred herb."

And that is a big part of the "set" when using it as MM.

It's whatever the patient wants it to be.

As my book you are holding in your hands was going to press, the editors of a respected journal of psychiatry honored him and his work by publishing a painting by the visionary artist Alex Grey of Sasha and his wife Ann on their cover and the lead articles on using psychedelic drugs to help heal, which I show on the next page. The painting portrays Sasha holding a generic psychedelic molecule that radiates light and healing.

Ann is holding it with him and gazing into the Light. [23]

---

22   Personal communication Esalen Institute, CA 1986.
23   Reproduced by kind permission of the Royal College of Psychiatrists to publish cover of BJP 206:1, granted 3 February 2015.

JANUARY 2015   VOL 206 NO 1

RC
PSYCH
PSYCHIATRISTS

The British Journal of Psychiatry

BJPsych

Making
a medicine
out of MDMA
Ben Sessa
& David Nutt

Efficacy of
cognitive bias
modification
interventions in
anxiety and
depression:
meta-analysis
Ioana Cristea et al

Positive
attributes in
children and
reduced risk
of future
psychopathology
Pablo Vidal-Ribas
et al

Recovery-focused
cognitive–
behavioural
therapy for
recent-onset
bipolar disorder:
randomised
controlled pilot
trial
Steven H. Jones et al

The March 1999 balanced
Institute of Medicine Report noted:

"Marijuana is not a completely benign substance. It is a powerful drug with a variety of effects. However, except for the harm associated with smoking, the adverse effects of marijuana use are within the range tolerated for other medications.* Thus, the safety issues associated with marijuana do not preclude some medical uses." Page 126-7

These authors do not appear to have been biased.

I see this quote applying most to the types of cannabis with a high percentage of THC and not for high CBD that has 0.3% or less THC. Most of the high CBD/low or no THC oil or tinctures are usually benign and are often healing.

*These 'other medications' include all of the countless drugs that were and are FDA approved. Many of them show a history of having substantial toxicity, including most of the psychiatric drugs available today, such as antipsychotics, lithium, and antidepressants.

Joy JE et al 1999 *Marijuana and Medicine*: Assessing the Science Base, Institute of Medicine, March 17, 15 845-1132

medicalmarijuana.procon.org/view.answers.php?
questionID=255

7 Using Medicinal Marijuana

Case History: Sarah

Sarah was in her early 70s and diagnosed with end stage breast cancer. It spread to other parts of her body and she was told she would die within six months. The physicians gave her the choice of chemotherapy or living as comfortably as she could for the time she had left. She chose chemotherapy because she had made up her mind that she was "going to fight this thing." And she did that in the most graceful way possible and her physicians were pleasantly surprised that she lived for another six years. This was in the 1970s when most people wouldn't have lived another year. She laughed and made fun of the fact that she had spent her whole life dieting to keep a slender figure and now she had to "force feed" herself to survive because the chemotherapy had caused her to develop cachexia. [24]

Once or twice a day, she took an inhalation of cannabis. She enjoyed the side effect of being hungry and she ate well. Each

---

24   Cachexia or wasting syndrome is loss of weight, muscle atrophy, fatigue, weakness, and significant loss of appetite in someone who is not actively trying to lose weight.

time I visited her she had a bag of Oreo cookies in her hands or nearby. I asked her if she was also eating nutritional food and she reassured me that she was. And with a big smile on her face she told me that her entire life she had craved Oreos and stayed away from them because of her constant dieting. Now, she considered the Oreos as the "fun part of having cancer."

First Time

If someone is sick and losing weight rapidly, they may be desperate and willing to try anything to get better. Many times they have never tried MM or recreational cannabis. The Institute of Medicine (IOM) reviewed the available data on marijuana's medical benefits and risks and concluded, "Nausea, appetite loss, pain and anxiety are all afflictions of wasting, and all can be mitigated by marijuana." [25] (IOM Text Boxes on pages 50 and 62)

It's common for some, especially first time users, to feel nothing the first time they try MM. They may have been using opiates for their pain which many times will make them more uncomfortable psychologically. They may not notice the effects of the MM because they are looking for a heavier feeling. Eventually, we together decide if they can cut back on the opiates because the MM is helping them with the pain management. If they have been taking opiates for over a week or two, if they taper the dose it should be done slowly, over weeks. This is the time to consult with the physician if they know to help taper.

When acting as a helper for a patient wanting to try MM, I usually ask them who makes them laugh the most. Many times I hear The Three Stooges, Laurel and Hardy, Charlie Chaplin, Monty

---

25   Joy JE et al (2003) *Marijuana and Medicine*: Accessing the Science Base. Institute of Medicine report. National Academy of Medicine, Wash. DC

Python and more. At times I may have a collection of these funny people on DVD and tape, just in case they don't have any. Notice I'm mentioning comedians who do physical comedy. This takes patients out of their heads where they spend a lot of time in fear – to watching something outside themselves that often times results in a good belly laugh. So skip the cerebral comedians who are funny too but force us to stay in our heads to get the joke.

For a first time user, when I ask if they had any feelings of being "high," they usually say no. But then they start talking about how funny something is and how good it feels to laugh. As Norman Cousins reported, laughter may be the best medicine. It lifts us out of our drama. I point that out as I suggest a snack which is hopefully (if the MM is working) accepted with zeal. Also offer fluids continuously. I would not include any sugary or alcoholic drinks.

I have not worked personally with the newly developed super strong varieties. I learned the experiential language of MM using its leaves and low potency early flowers and sometimes buds. That is mostly a bygone era. Today if you use the strong herbs I advise minute amounts, no more than a pinch of a bud. This is because it takes only a small bit if your set and setting are appropriate. People who go beyond the small amount necessary often become overwhelmed by the 'high' and miss the consciousness expanding level I am describing.

If the patient has been a user in the past, then they can set their own dose. Many people working at a dispensary usually can advise on the type to be used. Some of them may be used to advising mostly frequent and heavy recreational users who are tolerant and want only a strong experience. This may not be the best approach for MM patients. There is a difference between the dispensing staff sometimes called "Bud Keepers," when compared with more knowledgeable shop attendants. I suggest asking what strain would be the best for what the patient needs to achieve?

While each patient still has to use trial and error and attend to set and setting, Michael Backes' book *Cannabis Pharmacy* is one source. [26]

Tolerance for any psychoactive drug happens when the person has to use progressively increasing doses over time to get the same psychoactive effect. This physical and psychological occurrence happens commonly with regular cannabis users. The way to avoid tolerance and make any psychoactive drug work better, including cannabis, is to use it less often. This principle applies more to the THC effect than the CBD effect. If CBD (with no THC) is helping someone then using once or twice every day is an appropriate dose.

John Hergenrather MD said, "Although the 'high' [THC associated] is significantly diminished with regular use of cannabis, control of pain, nausea, anorexia, insomnia, and muscle spasm does not seem to diminish with frequent use. Dependence is not a significant problem with cannabis users. [Some disagree] If it is available, they prefer it to other pharmaceuticals." But some 9% become harmfully involved or addicted to it.

---

26   Backes M, 2014 *Cannabis Pharmacy*: The Practical Guide to Medical Marijuana. Black Dog & Leventhal Publishers New York, NY

If it is unavailable, they get by without withdrawal symptoms or dysfunctional behavior. There are those who admit to using too much, but no one that I am treating shows any classical signs of addiction, such as physical and psychological habituation to the use of the medicine, the deprivation of which gives rise to symptoms of distress, abstinence or withdrawal symptoms, and an irresistible compulsion to take the drug again, often in increasing amounts. Though the National Institute of Drug Abuse (NIDA) would have you believe otherwise, cannabis is not addictive, nor does it have a significant withdrawal syndrome." [27]

Sativa and Indica

Sativa strains of cannabis are regarded as having a lighter, more 'up' energetic high than *Indica* strains. Sativas typically have a more uplifting, spacey, cerebral, 'head'-centered high. Many medical marijuana patients report them as being highly effective against nausea, and sativas are generally effective appetite stimulators. Many experienced marijuana users prefer them for daytime smoking.

Indicas are known for their heavy, 'couchlock' body centered highs and are prized by patients who use them to help deal with chronic pain and insomnia. They can be a great aid to relaxation and make you sleepy. Using large amounts of indica thus often results in sedation and a nap.

---

27   Hergenrather J MD, 2014 *"Prescribing Cannabis in California"* ed. Holland J MD *The Pot Book*: A Complete Guide to Cannabis, Its role in medicine, politics, science, and culture. Park Street Press p.416-431.

Many experienced users prefer Indica for nighttime use. [28] Not all strains have these clear cut differentiating characteristics. Many have mixed features. So it comes back to the trial and error principle for any user, even with a rare laboratory analysis and report on a strain's chemical content. Trial and error still works best.

Table 2 Differences between Sativa and Indica *

Plant	Sativa	Indica
Leaf appears	Narrow	Wide
THC to CBD/CBN ratio	Higher	Lower
General effects	Stimulating, energizing, uplifting	Sedating, relaxing, grounding
Effect	Mental and emotional	Physical
User tends more	Extravert	Introvert
Best for	Day time	Night/bed time
Increases	Alertness	Sedation,
May help	Low energy	Fear/anxiety
May help also	Pain relief, nausea; a muscle relaxant	Insomnia, PTSD

  * *The degree of most of these effects depends on the dose.*
    See Appendix 6

Strains 101 How many strains are there? One source reported that 2,000 different strains of cannabis had been identified. What do we know about this practical but often uncertain variable of MM? Should the buyer beware? If so, when and why? *Summary*: We are learning as we go along.

---

28   Elliott S (2011-06-26). *The Little Black Book of Marijuana*: The Essential Guide to the World of Cannabis P. Pauper Press. Kindle

Leafly.com said they had found 779 strains, of which

218 were **Indica**, or **30%**

130 **Sativa**, or **17%,** and the rest

431 were **hybrids** or **55%** which was the most common 'strain'.

A savvy weedblog.com observer said "I've seen new strain names pulled out of thin air, which didn't come about from a new strain being created, but merely from someone taking a marijuana they bought and giving it a catchy name to help sales. For a long time I considered strain names to be unreliable at best, or a total sham at the worst."

Obviously, looking at Table 2 above, Indica would be the "plant of choice" for healing when someone needs to stay off their feet and rest for a set amount of time because it tends to sedate. If they are starting to feel anxious or afraid, I simply reassure them in plain language that the MM is going to help them. Too often their fear could be coming from "The War on Drugs" mindset, and now that war is over in half of the US States and several other countries.

Working with patients I make sure we have continuous eye contact and coach them on taking slow deep breaths starting to fill their lungs and going all the way down to their belly. Hold for a few seconds and then slowly releasing from their belly and moving up through their chest to their shoulders. If they need to slow down even more, I suggest we both pretend we have a straw in our mouth and breathe in and out through the straw. (This slows down the breath so there is no hyperventilation to "feed" the fear. Always be aware of the patient's breathing patterns because this is key to their being able to relax. Breathe slowly with them while having eye contact and holding their hand or rubbing a shoulder.)

When the war on drugs, especially on marijuana, was in full swing years ago, there was negative publicity about the side effects, paranoia, weight gain and lethargy. Now as a medicinal natural plant we hope for relaxation and (temporary lethargy is not bad here) it is allowing the person to relax and help them not to focus on their 'disease.' Weight loss is no longer a threat because the MM may stimulate appetite. The paranoia isn't necessary because of it now being legal.

Santa Maria

I have used this term four times. What does it mean? If the caregiver understands set and setting then there is a positive and healing spiritual aspect that emerges. It's something I can try to explain, but to be understood it must be grasped experientially through our hearts.

Twenty years ago in a shamanic ceremony, in the tradition of the Kahunas of Hawaii, I inhaled marijuana that was being called "Santa Maria" first in a ceremony of building an alter with my fellow participants in healing. Each one brought something sacred to place on this alter and it would be returned to us afterwards to take home knowing that it would then contain the vibration of our sacred ceremonies. Then we were given a brew of mint tea containing chopped-up psilocybin mushrooms. The Santa Maria expanded my senses to the point where, as I swallowed the tea, I felt the spirit of the mushrooms fill me and my energy field. It infused me with its energy in a way that I sensed I was under the influence of both of these sacred plants. We prayed before each ceremony and we prayed in gratitude at the end of the ceremonies.

I have seen and experienced that prayer works.

Giving the marijuana the sacred name of 'Santa Maria' and calling the psilocybin "food of and from the Gods" contributed to a mindset that I can still call upon when I drink plain mint tea. (This may be a placebo effect but the set and setting of the shamanic ceremony was so powerful that my physical body responds to mint tea in this moment based on that memory.)

There was and still may be a specific strain of Marijuana sold in Amsterdam that is called "Santa Maria." This is not what I am referring to here. Any strain that is prayed over or even asked to help someone is what initiates the intention for the plant to become Santa Maria.

Set

I am open to the gifts of these plants that are here with us on Earth for our benefit. As Sasha Shulgin had told me all those years ago, "You hold the marijuana in the palm of your hand looking at it. You ask it, 'Can you be this for me?' [Fill in your wish] Whatever this 'this' is, let the patient decide what he wants from this sacred herb." And that is a big part of the "set" when using it as MM.

It's whatever the patient or us anticipates, expects or wants it to be.

Set also means other aspects of our mind and our emotional set as our current worries, feelings and mood.

Whether you are the patient taking in MM or you are the helper "sitting" with the patient, the "set" is important. Only safe people should be in the room when MM is initiated. Once the patient is used to it and feels comfortable, then with their permission others may come in.

I have mentioned the role of the sitter throughout this book. As a sitter we have no judgment and no agenda. We can stay quiet until the patient asks us something. As a patient, part of the set is to relax and breathe the first breaths deeply and slowly about 3 times and then let it resume naturally. If we start to feel tense, the breathing should be noticed again and repeated as above. Coming back to our breath occasionally keeps us coming back to our self and our own inner life.

More often than not, either in the middle or the end of these sessions I am describing here, there will be laughter, which is good. Laughter discharges or moves blocked energy, as do tears. Do not be concerned over "fits" of laughter. They will quiet down, but laughing as we are getting in touch with our own inner life is a release mechanism. There also may be some moments that appear to look as though a "breakdown" is happening. If the patient and the sitter can allow for this kind of breakdown moment, it can evolve into a breakthrough that is positive and cleansing. When I am the sitter, the only time I will intervene is if I hear the patient trying to deny, repress or ignore their situation. I do it gently, and if I still get resistance, I back off, with compassion.

And remember to have:
No expectations.
And
Remain fully aware and anticipate **positive results.**

Eliminating Pain and Suffering

Both sativa and indica work well for *pain and suffering*.
Physical pain comes from our physical body's experience.

Suffering comes from what we are telling ourself about our pain that we perceive in our inner life:

"This is terrible!"

"I don't deserve this!"

"Why me?"

"I can't take this anymore."

Suffering also comes when we resist what is. Most times if we face it and deal with it, we will stop suffering about it. [29]

Here is a good opportunity to let MM help us to stop the suffering:

—Observe these thoughts going through your inner life.

—Now, put something that makes you laugh on your TV, computer, etc

—After several good laughs, go back to your inner life and see if you are still hearing suffering thoughts?

—If you are, then add some new lines like:

"When I laugh, I feel better."

---

29   Suffering comes when we are in conflict and don't know what to do with it. Two books that describe different levels of conflict and how to handle them are my book AFGEs and Charles Whitfield's book on Core Issues called *Wisdom to Know the Difference*. Reading and understanding what conflict is about can help us to address it with or without using MM.

"That's my ego trying to distract me from my life."

"I may hurt, but I don't have to suffer."

I'd rather watch a funny _____ then listen to this stuff in my head.

Marijuana Use and Death

How much should a new cannabis user worry about dying from first time or later use? Here are some facts. The annual causes of deaths in USA were found to be from the following:

Smoking tobacco 435,000 deaths each year,

Poor diet/no exercise 365,000,

Alcohol excess 85,000,

Legal prescription drugs 32,000,

Car/truck crashes 26,347,

Homicide 20,308.

For cannabis use directly there were zero. None. [30]

But indirectly cannabis-associated deaths, e.g., from an accident or uncaused event while intoxicated, there were a total of 279.

Most of these were determined to have been from a cardiovascular incident such as a heart attack (also called an acute myocardial infarction).

By contrast, the total number of yearly deaths from a few selected FDA approved and commonly prescribed drugs was 11,687. [30]

---

30   medicalmarijuana.procon.org/view.resource.php?resourcID=000145

My Plan

After assisting countless people make their transition to a non-physical reality, I have my own transition plan. Here is my idea of the set and setting.

I am at home in my own bed. Either alone or with safe others that for me would include my husband and our adult children, we could pray over the MM and ask it to be Santa Maria.

I don't want to be "numbed out" on opiates or other drugs. I want the best Indica available. I want beautiful music, probably Constance Demby (Gateway is her best), mixed with Elvis, (for lighter moments), some Beethoven and of course, Mozart! I want my kids, grand kids (and I hope I'm still around for) my great grand kids to visit interspersed with plenty of time in between for naps.

My "set and setting" is one that encourages memories of my life with whoever wants to share our memories. Resting interspersed with visits and hopefully laughter as well as tears are invited because I want and need everyone to be real and honor whatever is coming up for them as well as honoring what is coming up for me. I've had an amazing life. And I expect it will be amazing in the next one too.

No funeral, Please! Anyone who is interested can throw a good-bye party and somewhere during it – Give thanks for me to who or whatever we call "Our Higher Power." Please make this party a ritual of "Celebration." I plan on continuing this journey to the non-physical reality full of gratitude, humility and all the love I gave and got in this lifetime. I would like to be able to focus on this love as I leave because for me it is God in Action.

And of course, as Steve Elliot so eloquently wrote: "Euphoria and brilliant storms of laughter; ecstatic reveries and extensions of one's personality on several simultaneous planes are to be complacently expected." [31]

---

31   Elliott S (2011-06-26). *The Little Black Book of Marijuana*: The Essential Guide to the World of Cannabis (Kindle Locations 919-920).

8 Using Medicinal Marijuana
Continued

Setting

Setting is a safe environment. Ideally, that is.

Usually it's the patient's home, even their bedroom if there are children playing somewhere else in the house. Or it can be any safe place where the patient feels comfortable. Also, it is important that family members that the patient doesn't feel safe with – are not present. (I mention this key point about safe and unsafe others also in the hands-on meditation in the Appendix at the end of this book.)

I always let the patient lead in what we may talk about.

I have no agenda other than letting them talk about whatever comes up. And sometimes that means using headphones so they can listen to music, or a movie, or even them needing some private space for themselves. But mostly people want to talk to someone they know is open to hear whatever. They should know we are compassionate, non-judgmental, and don't have the same baggage with them that their family does. This key principle applies to and in all get togethers where any psychoactive drugs are being used in any circumstance.

If this is your family member you are attending to, and there is baggage, be ready to hear things you may not want to hear. And be able to let it go. That may be hard at the moment, but my experience is that later on you will be grateful that you handled it that way.

Sometimes, in my own inner life, I get urges to ask or suggest something. Mostly this is about asking if the patient wants to say a prayer or I can say it about something they are talking about. Do this especially if they are verbalizing a fear. I start by asking if they would be willing for us to put this fear into a prayer so they can release it. If they say, "Yes," I start this prayer (with their name for whomever they pray to) and they will chime in to add, correct or spell it out themselves.

They always lead and we always follow. It's about them, their situation, or if they request (or we strongly sense) about dodging the situation with funny videos, family pictures, psychobabble for the fun of it.

Whatever comes up for them. It is their sacred time.

Sometimes, when people are in more emotional pain than usual, this may not signal a breakdown. It can be the beginning of a breakthrough. Don't try to talk them out of their pain. Just listen while they express it. We don't need to help them work it all through. We just need to be there for them so they know they are being heard. If this happens often and you are feeling exhausted being a good listener, look for others who are safe to take turns "team" listening.

If you intuit the need to, reassure the patient that this isn't the

MM talking. It is them. The MM is helping them to relax, to eat and to let things come up that need to from deep within them. Or if they are the type that doesn't like to open up their own inner life, if it feels right, reassure them that you want to hear what is coming up inside them. By inviting it to become Medicinal, doing that keeps us in the Now. Praying over the plant while it is growing, praying before ingesting it, asking it to be Santa Maria, or simply intending to feel better because you or the patient knows it's going to help – is all that may be needed to enter into the Mystical Here and Now.

When we are in the Now, we can remember that God/ Higher Power/Spirit is everywhere. This experience is nothing we can grasp with our minds. But our heart knows it when we open to the Mystery that these plants can reveal to us.

As I reread what I just wrote, I believe I could sum this up for caregivers and loved ones as:

Listen to your heart.

Lead with your heart by listening more than leading.

Or as a colleague, Sister Girard, said:

"Don't just do something! Stand there!" [32]

If you have concerns about being capable of sitting with a patient, so do I. Every time I see someone who is very sick and possibly dying, before I walk in, usually on the drive over, I pray

---

32   Whitfield CL Springfield IL January 1976, personal communication

because I hear my ego trying to get me trapped in its doubts. I ask to be a conduit or a channel of Higher Power (or whatever you are comfortable calling the God of your understanding). I ask that ITs wisdom and love come through me to whoever I am visiting. And I ask for help getting my ego out of the way (any of my 'agendas') so that IT may come through. And of course, when I leave, I say a prayer of gratitude. These two prayers, before and after, take the worry away. They connect me to our Higher Source, and seems to connect me to the Soul of the plant that is coming in with its healing and spiritual properties.

And please be reassured that no one has ever died from using marijuana. If they want to take more than you think is necessary, they're in charge of it. There are so few things left in their life that they are in charge of – let them be in the driver's seat on this one. The worst that may happen is they will eat too much or fall asleep.

Case History: Sherry

Sherry was in her early 40s. I was called in to her case three days before she died. Her still young body was filled with cancer. Her attitude was to protect herself in her final days from her family. She let her husband and best friend come into the bedroom. There were at least ten other people milling around the house, wanting to see her and knowing they couldn't.

I immediately realized that her pain medication couldn't work because she was blocked, and a few minutes after swallowing morphine she vomited it back up.

I called hospice and they immediately changed her meds to a fentanyl patch. She knew she couldn't eat anymore but she was still able to inhale. We initiated small doses of MM for the nausea and vomiting and it calmed her down as well.

She told me dying was boring and I agreed. Then she told me she had been sexually abused by her mother and her grandmother as a child. They were both dead but the rest of the family, the ones milling around outside her door had since long ago never stopped them from hurting her.

She asked me to put on the television and I started moving through the channels looking for something light and funny. Suddenly Sally Field was standing over her daughter, Julia Robert's grave. Sally Field gave an Academy Award winning performance standing over that grave. Her daughter had just died and this was the end of the funeral. The other women, the 'Steel Magnolias,' just stood there witnessing and allowing. When Sally Field finally stopped, someone said something funny. And with the intense cleansing laughter, they left the grave. I tried earlier to change the channel but Sherry wouldn't let me. I was crying and she asked me why. I told her my dad had died a few months earlier and we never had a chance to say what we needed to say to each other.

Next she asked me to put on the VHS tape that was sitting on top of the TV. It was labeled "Sherry's birthdays." And there before our eyes was a yearly party given to her all through her childhood. She asked me what I saw and I told her there were many adults that looked at her with love. I saw lots of hugs and some genuine kisses.

Sherry fell asleep after that and I left to go home and get myself back. When we are doing this kind of intense supportive work, we need to have breaks because we or at least I leave myself at the door and are totally there for the patient. If we don't get breaks, especially loved ones who want to be there every second, we develop what is called "Compassion Fatigue." If the caregiver can't leave, if there is no one to relieve them, there are support groups for caregivers on line. Your local hospice or cancer society can give you the link.

When I came back a few hours later, Sherry was sitting up in bed and all the people in the house were now sitting circling her and smiling. There were a lot of stories shared and some laughter including Sherry. About an hour after I came back she slipped into a sound sleep and died about an hour after that.

When the hospice nurse came a little while later to pronounce her dead and fill out the necessary papers, she asked me what had transpired. She had been visiting Sherry every day and said there was a huge difference in this final visit. With that question I wrote a book of 14 stories of people I have sat with and helped die. It is called Final Passage: Sharing the Journey as This Life Ends.

What to eat and what not to eat

Dean Ornish MD summarized how food and lifestyle habits can increase brain cell health. Increasing brain cells: Cannabinoids, blueberries, tea, moderate alcohol, stress management, chocolate. Decreasing brain cells: Sugar, nicotine, excess alcohol, chronic stress, cocaine, opiate painkillers, saturated fats.

Cannabinoids have been shown to promote embryonic and adult hippocampus neurogenesis and produce anxiolytic and antidepressant-like effects. [33] CBD is a natural blood and tissue anandamide level stimulating chemical that also blocks hypoxic and other stress related glutamate toxicity and is neuroprotective. [33a]

Sugar danger There are numerous anecdotal stories that sugar (including high-fructose corn syrup, 'brown' sugar and honey) contributes to cancer, diabetes and

The American Cancer Society writes on their website:

'Sugar increases calorie intake without providing any of the nutrients that reduce cancer risk. By promoting obesity and elevating insulin levels, high sugar intake may indirectly increase cancer risk. White (refined) sugar is no different from brown (unrefined) sugar or honey with regard to their effects on body weight or insulin. Limiting foods such as cakes, candy, cookies, sweetened cereals, and high-sugar beverages such as soda can help reduce sugar intake.'

Theories supporting limiting sugar state that sugar creates more circulation of arterial blood to the tumors. Studies are being conducted now on CBDs' (see Chapter 11 on CBDs) stopping cancerous tumors from growing by cutting off circulation to the tumor.

If the patient has any cancer at any stage it's a "no-brainer" to stop eating sugar and that includes any prepared food that has sugar in it. Sugar weakens our immune system. Eat fresh foods only and if it's available—real organic.

---

33   Jiang W. et al. 2005 *Journal of Clinical Investigation*, 115, 3104-3116.
33a   Leweke FM et al CBD enhanses anandamide and alleviates psychotic symptoms. *Translational Psychiatry* 2012 2,e94 (Nature.com)

That may mean more time preparing meals but it will be worth it. Fresh fruit and vegetables should be available to the patient around the clock to help with (hopefully) the munchies. Hummus and Guacamole can be made fresh (with no sugars or preservatives) to dip the veggies in and add more nutrition. Research for false 'organic' food labels that sometimes are labeled 'natural.'

Case History: Dave

While Dave's story does not directly involve MM, it does address the importance of nutrition and sugar. Dave and his wife have been dealing with his "terminal" brain cancer for the past year. He did one round of chemotherapy and then went on a "Ketogenic" diet. He has lost 90 pounds and his wife lost 50. He also walks 3 miles a day. He recently had an MRI. His tumors are gone. Two small cysts remain but are not cancerous. The doctors now believe that they will reabsorb in time and he is going to live. Dave's wife told me:

"Dave (and I just because I do the cooking) started the ketogenic diet the morning after he got home from the hospital [a year ago]. My best friend was researching it all along and came over with the research the night he got home. The doctors were amazed at how well he tolerated both the radiation (1 long and 1 short round) and 1 round of chemo over a year ago. His oncologist expected his tumor to look worse immediately after radiation, but Dave's looked better! He said "Must be that diet you're on!" That was over a year ago. There is some research that shows that people do better with radiation and chemo side-effects when they are on the ketogenic diet, too. I mean the brain is how much fat? Lots." [34]

---

34   Ketogenic: [If you or a loved one has cancer and is thinking of going on this diet, medical supervision is recommended.) The diet is high in fat, supplies adequate protein and is low in carbohydrates. ... continued on bottom next page

Children, MM and other drugs

The best place for a patient to use MM is in their home. If a patient has children living in the home, the MM should be locked in a safe place.

MM is not for children to try and be careful to place it beyond their reach if it is in a food product. Children eating any marijuana can become extremely ill with disorientation, rapid heart rate and fears, anxiety or over sedation. I am also a strong believer in no exposure to marijuana until a person is fully grown and mentally healthy with an appropriate amount of self-confidence. This confidence would help prevent mental instability in the face of inability to handle conflicts on all levels.

My strong beliefs are not likely to stop any teenager who wants to experiment with drugs. And because for so many years our government dictated drug policy that insisted on no drugs, teenagers don't believe the warnings they are given. What we need in our world today are real facts about what drugs do to our body, mind and spirit. Hopefully, with this new movement across the USA and Canada, marijuana will no longer be categorized with heroin, cocaine, or the psychedelics such as LSD, ketamine and others. Marijuana is no more likely a "Gateway" drug than nicotine, alcohol or prescribed psychiatric drugs. Some have observed that MM can be an "Exit" drug.

...This combination changes the way energy is used in the body. Fat is converted in the liver into fatty acids and ketone bodies. Another effect of the diet is that it lowers glucose levels and improves insulin resistance. An elevated level of ketone bodies in the blood, a state known as ketosis, leads to a reduction in the occurrence of epileptic seizures. For more information: Go online to charliefoundation. org/explore-ketogenic-diet/explore-1/ introducing-the-diet

A small percentage of teenagers become psychologically addicted but usually there were already psychological problems. Teens should be warned that if they are going to try occasional use, that it not be during school, while operating a car or any power equipment, nor while any responsibility is expected of them.

Young brains exposed to marijuana may not develop normally. That also goes for alcohol or uppers, downers, opiates and psychedelics.

Our children watch us have a glass of wine with dinner.
When we are out we may even have a cocktail in front of them. But healthy parents do not get drunk in front of their kids. They show them what responsible drinking looks like so they have an idea of what they will emulate. The day may come when we will be doing the same with MM or recreational marijuana if it becomes legalized.

If you suspect that your child or teenager is using marijuana or other drugs, here are signs to look for:

A sudden change from their usual personality or demeanor

Changing groups of friends

Sudden drop in grades

Memory impairment

Difficulty paying attention or solving problems.

Paranoia and delusional behavior

Overspending money

Sudden increased appetite

Red eyes.

9 Cannabis Delivery Methods

F or first time users:

Be careful if you don't know the strength of the MM used in any of the following ways, as otherwise you could overdose and have an upsetting experience. If in doubt, start slow and at low dose. Be sure you have a comfortable set and setting.

Smoking: Most people prefer to smoke cannabis for several reasons. These include: a rapid onset of effect, a familiar method, convenience, easy dose control, cost-effectiveness, and for some patients, a reliable medical effect or symptom response. Others may prefer vaporizers to inhale the heated plant material, including its waxy resins, without actually burning it. To do that they use electronic devices to heat the material to baking temperatures of from 290 to 446° (degrees) Fahrenheit (180 to 230° centigrade). [35] But realize that as is true for beginners using the cannabis plant, these devices also require trial and error.

I summarize the characteristics of the various methods in Table 3 on the next page.

---

[35]   Hazecamp, Pappas; If interested, do a net search for more information.

Table 3 Nine MM Delivery Systems[36]

Delivery system	Efficacy	Onset (minutes)	Duration (hours)
Smoking: pipe, water pipe, 'joint'	Most used over time	1 to a few.

Vaping may take a few min longer than smoking | 1 to 2+ |
| **Vaping**: with an electronic vaporizer that heats herb or 'oil' | More subtle onset

May need less MM | | |
Ingesting *in food* – know potency per dose	Can overdose due to slow onset	30 - 90	5 to 8
Tea	Excellent so no part of the MM is wasted	Immediate to 30	Long, lower side effects
Patch	Still in research. Looks promising	No data	No data
Drops, Liquid: oil, tincture Some find vaping the same oil more economical	High CBD/ low to no THC usual	On tongue, fast acting	10 to 12
Sublingual **spray**	Varies. See pkg insert	5 – 20	Up to 8
Salve	Said good for painful joints but stickiness may linger	A few	2 or more
'Dabing' caution - this is *not* for beginners. Know potency per dose.	If over dosed can be overwhelming and disorienting.

Intense and within seconds | | May last longer than smoking or vaping |

36   Hergenrather J 2014. *'Prescribing Cannabis in California'* 36 *The Pot Book.* ed Holland J; I expanded also from California NORML

If the patient is a cigarette smoker and may get too sick to smoke cigarettes, they may want to smoke their MM instead of tobacco. But I advise my patients not to think they can substitute the same frequency of inhaling MM as they did for cigarettes. If that happens they will smoke much more than they need. They need to be guided to learn the difference between smoking far too much cannabis and learn how little they need.

Vaporizing: Depending on the details, the herb is still inhaled as vapor or smoke. THC vaporizes at a lower temperature than it burns. Some vaporizers heat cannabis to about 311° (degrees) f (155° c), below the burning point of a combustible plant. So the vapors are cooler, purer and probably less toxic than smoke. [37]

Some 'pen' vaporizers have a tiny metal coil at the bottom of the chamber that heats red hot that burns the herb to over 446° f causing enough smoke to irritate the throat and bronchi but is less irritating than a joint or pipe.

Although 'vaping' is the closest experience to smoking and shares some of its immediate effects, at the above low temperatures it does not fully combust the plant material, and most are considered safer than ordinary smoking for the lungs and respiratory system. Vaporization has gained popularity, especially within the medical marijuana community, as a healthier alternative to previous smoking. This method also conserves marijuana since less is typically used. [38]

A more efficient way than regular smoking is to combine liquid glycerol with a few rice grain size bits of wax or dabs in a tank which takes a day or so at body temperature to promote mixing before use.

---

37   Dale Gieringer, California NORML Director

38   Elliott S (2011-06-26). *The Little Black Book of Marijuana*: The Essential Guide to the World of Cannabis (Kindle Locations 838-839). Peter Pauper Press. Kindle Edition. Ref 35 above

Some prefer vaping at lower heat for the gentler effects on the lungs. [39]

Inhaling **CBD in oil** in a simple CE5 or Kanger protank mini *tank* device with an inexpensive Vision Spinner II *battery* is usually more effective and economical for problems as PTSD, insomnia, anxiety and drug withdrawal. Many prefer this method over smoking raw cannabis in a bowl or 'joint'.

There are hundreds of products now online claiming to vape the herb. Some produce regular or slightly visible smoke and others make no smoke. When inhaling and exhaling with the lower temp devices users tend to notice only a mild taste of the herb. Wait a few minutes before trying another dose. It may seem as though it can't work if there is no smoke, but after a few minutes the patient may realize that they are feeling the effects that, although subtle, they are different than how they felt before they inhaled it. This use can be a first step in expanding our awareness. We start with what our physical body, especially our usual awareness of it, feels like. Then we can move on to our inner life and what we are experiencing inside.

Sublingual Spray: This method may have some advantages. An atomizer device dispenses a dose of liquid cannabis tincture under the tongue. Absorption through the oral mucosa into the bloodstream. The user knows within 5-10 minutes if the initial dose was enough to lessen symptoms. The effects peak in less than an hour and last for 1 to 4 hours or longer.

The oral spray is not processed by the digestive system and liver into a metabolite as occurs with orally ingested cannabinoids. (NB: Vaping the same spray with zero THC but high CBD oil [not an alcohol containing tincture] via a CE5, Kanger or Aspire tank works faster and at less cost for many users.)

---

39   Hergenrather J 2014. 'Prescribing Cannabis in California' 36 *The Pot Book*. ed Holland J; I expanded also from California NORML

Using the sublingual spray may have a similar effect to smoking cannabis but without the smoke. Spray makers recommend starting with 1 or 2 sprays, increasing to 4 sprays if needed to control symptoms. As with all of the other delivery methods, the user is responsible for selecting the most beneficial THC/CBD ratio and dose for the symptoms to be addressed.

The pioneering British GW Pharmaceuticals firm markets sublingual sprays under the brand names of Epidiolex for epilepsy and Sativex (generic name nabiximols) for the pain of multiple sclerosis and spasticity. Sprays are also available by several generic makers. Generic sprays are listed on the Project CBD website at projectcbd.org/? s=spray. For further information on sprays see Appendix 6, under the cannabidiol discussion section on pages 162 through 164.

Ingesting in Food: Preparing MM in food is a whole book in itself. And there are several cook books out now that explain the way to prepare the marijuana before it is added to food, and then good recipes to make it palatable. You can't just fold it into a recipe. It needs attention, focus and planning before preparing it. I have seen it simmered in a stick of butter and then refrigerated. Then the green leaves are easy to skim off when the butter hardens. The butter will contain the important active ingredients. Some prefer leaving the leaves in the butter for nutritional and other holistic reasons. Remember to say a prayer over the "brew" to infuse it with good wishes and healing properties.

As a respiratory therapist I have treated many patients of MM in food because this is the safest for people with respiratory problems.

Use edibles with THC with extreme caution. One cookie or other food may have four strong doses. Try eating one fourth or less of one. If you feel nothing, wait at least 2 hours before eating more. Edible MM lasts 5+ hours. It is safer to wait 5 hours before taking a "booster." First learn how your body responds. Until you know the terrain, have someone safe overseeing your eating it.

Used carefully, ingesting the MM creates a longer and smoother effect. Eat it on an empty stomach and realize that it usually has a different quality and feels different than smoking it. It is subtle, smooth and can last 5 or more hours. It becomes more subtle after the 4 to 5 hour mark but it still has qualities that are rewarding if meditating or just relaxing. If eaten about an hour before lunch time, you may want to postpone eating. This is a ripe opportunity to learn to listen to your body.

Make lunch a light and nutritional meal. This state of consciousness will last until dinner and if you or your patient are bedridden or house bound, then another dose can be given before dinner with the same directions and effect. If the patient is ambulatory, a late afternoon accompanied walk in nature (or just around the block) could be a positive experience.

Drops, Oil, Tincture: Some prefer taking MM as CBD in glycerol 'oil' or tincture form, under or on their tongue for 1 1/2 minutes or massaging it into their skin.
Vaping it in a simple inexpensive CE5, Kanger, or Aspire tank and Vision Spinner II battery works better. The usual dose is once or twice a day and helps many with PTSD.

Tea: Drinking it as a tea has become popular and practical for some users. If growing our own plants, then we save the left over seeds, stalks and twigs, even the leaves that look dried out and not green like the usable parts of the plant, which can be brewed into tea. After soaking the plant material with boiling water and steeping for 15 minutes, strain off the twigs, etc. Leaf particles can be left in the tea for nutrition.

For those patients who can't smoke because of respiratory problems or eat MM because of digestive problems, trans-dermal patches have just been released.

'Dabing' This is a recent and controversial delivery method of inhaling the specially made wax from the plant and its concentrated trichomes for several reasons: 1) it is complicated to use (search online, including on You tube), 2) its safety varies by which chemicals were used to make it and 3) it is quite potent and thus not for a first time cannabis user. It is a kind of refined hash that usually deletes most cannabis' components except the THC and some CBDs.

Variations Using Oral Cannabis

Some people like to eat it and many report that it is a recognizably different and usually a pleasant experience. The onset starts in about 40 minutes to an hour. It lasts much longer and some report an effect of 5 to 8 hours or more. They say that it is smoother and some feel a sense of a more open heart. When using any of these delivery methods, low doses instead of higher doses will be associated with the ability to function better. Some find the ability to function better when they eat it. Chances of becoming anxious or paranoid are usually lower.

But the same wise precautions apply for all THC-containing cannabis uses: do not drive or operate anything mechanical that may be risky.

There are a wide range of responses to oral cannabis, and different patients given the same amount of oral cannabis will each absorb it at different rates and metabolize it with varying efficiencies and effects.

Some patients cannot tolerate MM near bedtime as it interrupts their sleep cycle. It may help some fall asleep, but some report that they can't stay asleep. This problem can be alleviated by making the last smoked dose at least 4 hours before bedtime. If the patient is at a chronic stage of any illness that MM may potentially help (and not an acute life threatening one), then one or two doses a day at about noon and again in the early evening may be effective.

With the increasing availability of cannabis product varieties that contain cannabinoids other than THC (such as CBD), oral cannabis preparations can be made with modified, reduced, or no psychoactivity—depending on the ratio of CBD to THC in the preparation. An 8:1 ratio of CBD to THC typically eliminates THC psychoactivity altogether. A 3:2 CBD to THC ratio will have some psychoactivity but it is usually more clearheaded. Trial and error here is again the principle, watchword and maxim.

These ranges are noteworthy because patients consistently note that these ratios are effective for reducing anxiety. Alternative cannabinoids, such as CBD and THCV, may also be of interest to patients for whom conventional THC psychoactivity may be a problem. Many experts consider our understanding of these alternative cannabinoids to be refined as herbal cannabis medicine evolves. [40]

---

40   Backes M 2014 *Cannabis Pharmacy*. Several sections on dosing information

Healing vs. Curing

When I refer to "healing" throughout this book, I am not referring to a "cure." With a cure, all symptoms disappear and the patient finds a new healthy balance that will help prevent another group of symptoms. I consider a healing to be both a lessening of the intensity of and a learning about the nature of our symptoms so that we understand them and there can be a new balance for the patient of their body, mind and Spirit.

Some of our symptoms and sense of dis-ease may continue, but we will now have a better understanding of what we need to do (and not do). Our inner work while using MM and reflecting on it later shows us new possibilities, including the need for stress reduction.

If we are the sitter, neither of these above outcomes, healing or cure, are our goals. Our only goal is to relax ourselves and let the patient be, say or do what they want and need to do.

Paradoxical or Opposite Reactions

As for nearly all medicinal healing aids, some people will develop the opposite reaction to the one that they wanted, also called a 'paradoxical' reaction. An example would be when MM is used to help reduce pain; some people feel their pain increase because of their having a heightened sense perception. Others feel that their pain hasn't decreased but the MM distracts them from focusing on their pain to other things, including music, taste, and even enjoying sex. Several patients reported that they were able to reduce the dose of their opiate pain drugs but not stop completely when they used MM.

To do that takes special medical expertise.

Using MM with high THC content alters our short term and sometimes our long term memory. Short term or working memory is the ability to transiently hold and process information for reasoning, comprehension and learning, such as active thinking. We tend to forget more often using MM than we do in our ordinary state of consciousness. [41] It can be humorous to suddenly forget in the middle of a sentence what we were saying. Or it can be frustrating if we thought we had a good point. I suggest having paper and pen in close reach and jot down the ideas that you may intuitively know you are going to want to remember later. Consider reading the footnotes on memory below carefully.

Witness Meditation

(The following meditation can be done with or without using MM. Try it each way and see which one or both works for you. But do not try this if you have just had a dose of MM and are "high." If you have used and are feeling its usual effects wait

---

41   1) theguardian.com/science/neurophilosophy/2012/ mar/02/how-marijuana-impairs-memory 2) Independent Drug Monitoring Unit accessed 30 Oct 2014 idmu.co.uk/canna-bis-memory.htm **Acute cannabis intoxication** disrupts short-term memory by interfering with the *filtering* of **information**, such that a *greater volume* of information **reaches consciousness**, overwhelming the ability to store, or prioritize such information for storage in memory. Acute cannabis intoxication causes dose-related deficits in most aspects of memory including encoding, storage and retrieval. Such memory deficits may be offset where the cannabis used contains significant levels of CBD. This effect is well established and research into the endocannabinoid system is currently leading recent developments in understanding of memory processes.

...Few users report severe memory problems. The evidence suggests that in the majority of cases, memory deficits do not persist during long periods of abstinence from the drug. Long-term users can learn to mitigate or decrease the deficits by novel strategies and by using different brain pathways to encode information. The few studies on long-term episodic memory find only weak associations between cannabis use and long-term memory deficits, which tend to include forgetting (particularly in cases of post-traumatic stress). [See Appendix 13, page 187 for more on THC's effect.]

until you are feeling like it is beginning to wear off. Make sure you haven't had any coffee, other caffeine products or sugar for the past two or more hours. Have a safe person read this to you if need.)

'Ask MM what you want it to be for you or say a prayer of intention. Then spend at least 20 minutes or longer meditating. Either sitting in a comfortable chair or lying down in a favorite place, take several slow deep breaths. If you are not a regular meditator, close your eyes and have someone safe read this to you: 'Starting at the toes, slowly go up your entire body and feel around for any tension. Ask it to let go and keep breathing slowly and deeply. If you need to, go over your body again. Ask your scalp to relax and then the small muscles around your eyes, mouth and jaw. Remember that you can also purse your lips like you have a straw in your mouth and breathe that way while realizing that the tension and stress are flowing out through the pursed lips too.'

'Turn your focus to your inner life. What you are hearing in your mind? What kind of monologue is going on as you relax? Is it fearful or painful? Does it sound like gloom and doom?'

Who is doing the observing?

'Step back from this chatter and continue to observe it.

'(Do not try to negotiate with the chatter, which will only energize the discussion and you will then lose your observer in a closed loop discussion, sort of like a hamster running on the wheel in its cage. We can't talk the chatter in our minds out of chattering.)

'You may want to pause now and open your eyes and come back to this book.'

*　　　　　*　　　　　*

The real you (real or true self) is the "witness" who is doing the observing, not the chatter in your head. The chatter (false self or ego) is part of the programing that started when you were very little. It is broadcasting all of your unfinished beliefs around your past traumas and linking them up to your current situation. This experience can possibly make you feel like a victim.

As you learn more easily and faster to move away from the chatter and Identify with your Observer/Witness self, do the pursed lip breaths a couple of times to blow off the tension that was building up from chatter. This breathing is cleansing and each time you feel you are blowing off the tension – you are learning more about your own ability to be in command of your body and mind.

The more you do this exercise where you stand back from the chatter and don't identify with it by just observing that something besides the voice of the chatter is there watching, the stronger your observer/ witness will become and the sooner you can learn to identify with it and not with the chatter. One more observation – when you identify with the voice of the chatter you may feel anxious, fearful or just overall angst. When you identify with the observer you will feel calm, even peaceful.

See Appendix 9 page 174 Voice of the ego vs The Real Self

10 Post-Traumatic Stress Disorder (PTSD) and Simple Stress

The *Colorado Springs Gazette* published this in 2014

Free Cannabis Giveaway [42]
"About 1,000 show up in Colorado Springs for veterans' marijuana giveaway."

A free cannabis giveaway at a Colorado Springs hotel Saturday attracted about a thousand people looking for an alternative medication for their physical and mental pain.

Roger Martin, the executive director and co-founder of Operation Grow4Vets, which put on the event, said the group's goal is to bring cannabis to veterans with service-related conditions as an alternative to pain medications.

"It isn't going to hurt them as much as the prescription drugs," he said.

Martin, an Army veteran, said he struggled with prescription drug use to help with what he called "24- hour" pain and an inability to sleep. "I just need something to take the pain away during the day," he said.

---

42   The *Colorado Springs Gazette* Sept. 28, 2014 by Stephen Hobbs

Martin said he discovered edible marijuana as a way to reduce pain and help him sleep more, and he wants other veterans to have the same chance to address ailments.

Matt Kahl, a former Army specialist who works for Operation Grow4Vets as a director of horticulture, said using marijuana saved his life and reduced his dependency on pain medication.

Kahl said he was injured when serving in Afghanistan when he was thrown from a vehicle, causing a traumatic brain injury and hurting his spine and back. As part of his recovery, Kahl started taking more than a dozen pain medications per month. After a suggestion from a friend, Kahl started using marijuana to help with the pain. Now, he said, he is off all but two of his medications. "It doesn't make sense that our first line of defense is toxic medication," Kahl said.

He said marijuana use lessened his symptoms of hypervigilance and pain, and he moved to Colorado, "I would not be alive without this," Kahl said.

People who came to the hotel Saturday were given a bag of items that included cannabis oil, an edible chocolate bar and seeds to grow plants.

Our Psych Practice

Our psychology, psychiatry and addiction medicine practice brings us face to face most days with Post-Traumatic Stress Disorder (PTSD). Adults who were repeatedly traumatized as children or as adults often are suffering with undiagnosed PTSD and are often mis-labled with a "mental illness." They are re-traumatized every time they go to another clinician who doesn't understand PTSD. They are commonly mis-diagnosed and labeled with depression, bi-polar disorder, anxiety disorders, or many of the other labels in the DSM (Diagnostic and Statistical Manual of Psychiatry).

A main difference between symptoms for depression and PTSD is "suicidal ideation." To diagnose PTSD we don't have to be suicidal. [43] Clinicians who don't understand this may label the patient 'depressed' and write a prescription for an anti-depressant and send them on their way with no real help or worse. Anti-depressant drugs don't work well and are often toxic. Patient complaints are many and include:

"They don't work."

"They worked for a short time and have stopped."

"I put on 20 pounds." (We've seen 80 lbs each in 2 patients)

"I can't get off these pills."

"When I complained to my doctor he gave me another prescription to add to what I was already taking. Now I'm on two." ... I have heard four or more.

"I feel numb all the time"

---

43 Personal communication: Douglas Bremner MD; also Whitfield C 2003 *The Truth about Depression*. and 2004 *The Truth about Mental Illness*. Health Communications, Deerfield Beach, FL

If prescribing physicians understood the difference between major depression and PTSD perhaps there wouldn't be millions of anti-depressants prescribed every year in this country. PTSD responds best with group and individual psychotherapy, Emotional Freedom Technique (EFT), and MM as suggested in the article above. One of the few prescription drugs we use in our practice is buspirone, a mild anti-anxiety drug that is easy to stop. This effective drug has essentially no withdrawal. We may add phenytoin (brand name was Dilantin) for some with more advanced PTSD, an old but still effective anticonvulsant shown to help heal trauma-induced brain lesions.

I refer to PTSD as the "Clenching Disease." My patients with PTSD agree with this and can feel it. They clench to such a degree that blood supply can be cut to the most vulnerable area of their body. The clenching can appear as headaches because the scalp and even the muscles around the eyes and jaw are tense. Clenching affects the digestive system resulting in reflux and other painful manifestations.

Breathing is also disturbed. The diaphragm barely moves and results in shallow breathing and the inability to sigh. And because of this clenched breathing, not enough oxygen gets to the lungs to then be transferred throughout our body. Deep slow breathing and sighing are an important way to blow off stress. When the patient has gotten used to using MM, it is then important to get in touch with how they are breathing. Meditating on the breath every day is important. It is easier to let the chatter in our heads go or at least quiet down when we focus on our breath and our belly. If breathing appropriately, we should feel our belly going up and down as it fills and empties.

Prescription Drugs

When anti-depressants don't work, unaware physicians may try anti-psychotic drugs that are more powerfully mind numbing and now referred to by some as "a chemical lobotomy." The family thinks the patient is better because they aren't acting out as much. But inside their thinking and decision making is stifled.

It is others' and our observations that many of our Vets who come back from war and develop PTSD were also abused or repeatedly traumatized as children. This prior trauma primes or conditions them all the more.

PTSD Case History

A 64-year–old woman who presented with repeated childhood physical and emotional abuse and neglect has been in one of our psychotherapy groups for 3 years. She also had private sessions with me for EFT (see footnote page 52). She receives massages regularly, meditates daily and exercises most days of the week. She tells us often that she is grateful for the relief she has experienced during these 3 years and sometimes sadly mentions that she knows she will never be "normal" like someone who was never abused. The therapy group and she often express an almost constant "fight or flight" sense in their body and inner life. Our groups' "mantra" on this is "when PTSD raises its ugly head – remember to name it." [44]

---

44   In other words, as soon as we feel ourselves being triggered by some outside appearing "threat," tell ourselves or some safe nearby person that we have been triggered and what's going on isn't actually real. We are reacting now to something that hurt us long ago. Just realizing and naming this can help to slow down and eventually stop us reacting in this way. Our groups have shortened this into "When PTSD raises its ugly head – remember to name it." We don't have to explain ourselves to anyone. We just give "it" a name, such as PTSD or the name of our issue.

Someone told her about a new discovery that is a part of cannabis called cannabidiol (CBD). It lacks THC (or the amount of THC is minimal), so the component of the plant that produces the "high" (strong psychoactive effects) is not active or not there. CBDs are increasingly being given to treat children that have multiple seizures that are unresponsive to legal but toxic drugs, commonly with excellent results. CBD in 'oil' (as glycerol) is dropped on or under the tongue or vaped in a small CE 4, Kanger or Aspire tank once or twice a day.

This patient started taking CBD drops and then found vaping the oil worked better. She now feels the "normal" feeling state that she once thought was unreachable. She no longer feels like she is "walking on egg shells" that is common for PTSD.

She calls it "A missing link" to her recovery.

Since 2010 CBD has been sold as a "nutritional" or "dietary supplement." Of course, it is far more than that. It is a natural bioactive treatment that works. The bottle says to start out with half a dropper or 15 drops on or under the tongue. She is taking 5 drops (a quarter of an eye dropper) in the morning and 5 drops in the evening. The least amount that positively affects the patient is usually the best amount. It can be ordered online. She soon found by trial and error that vaping the CBD oil was useful, faster and cheaper.

Breakthroughs

When I lecture at conferences on spiritual experiences and how integration happens within them, I am often told by therapists and lay people that they themselves or their patients with PTSD are able to transcend their damaged belief systems.

They say they do that by transferring all the energy that they had previously devoted to their suffering, to doing repeated and usually long term safe inner work plus using other aids that help them to physically reduce the tension in their bodies, into what eventually becomes a breakthrough into a higher level of consciousness. That inner work can be accomplished by any of several time-proven healing methods (see Chapter Six). We have long seen this in our private practice. By moving into their inner life instead of running away from it, their continuous state of hyper-arousal, anxiety (fear), and isolation becomes instead an expanded level of consciousness where their memories no longer contain so much of the emotional charge. They can still tell their story, but without the emotional pain they once had.

Many people with PTSD who suffer with the survival defense mechanism of "dissociation" (separation of our attention from the reality of our inner and outer life) learn to have compassion for themselves even though they have been labeled in the old paradigm as "dissociative." This trait of dissociating generates shame because they had thought it was their fault. When they look back or are helped to look back on their traumas they realize that "leaving" the painful scene was a survival oriented gift helping them to temporarily detach from their seemingly life threatening experience. And somewhere in that realization is a new found spirituality that first manifests as compassion for themselves and what happened to them.

It appears as though there is something in the MM that helps them to remember the sense of spirit in them when they left each painful scene. CBD drops (without the "high" of the THC) appear to take away the repeated fear and anxiety underlying dissociation.

In this realization, their painful stuck emotional energy transcends to a new felt sense of compassion.

... More on PTSD in the next chapter.

Simple Stress

De-stressing Exercise

If this feels appropriate (both patient and caregiver are agreeable) set a time to sit and talk about the stressors in both and each of your lives.

> No Judging.
> No solutions or advice.
> Just listen.
> Don't try to rescue.

The only response is simple listening without interruption; as well as Compassion for the other and for Self.

1. Take turns with who or what is upsetting you and why.

2. Take turns with why what is happening is upsetting.

3. Just saying anything that is stressing you out loud will help.

4. Do not take responsibility for each other's stress.

5. Agree to give yourself compassion around your experience that felt stressful. You know where it is coming from, so allow yourself and each other to surround your stress and its roots with understanding. Ask for it to slow down, get much smaller and take breaks.

Stress Reducing Meditation on Self Pity or Self-Hatred vs. Self Compassion

The MM caregiver can read to the patient from the next page out loud while the patient is either lying down or sitting in a comfortable chair, with soft music playing in the background. The reader should pause whenever it feels right. This can be done with or without MM.

Table 4. Effective Stress-Reducing Tips & Tools [45]

Adjust your expectations	Learn how to say "no" and delegate	Express feelings instead of bottling them up
Avoid stressful people and situations	Be flexible and willing to compromise	Pick your battles carefully
Focus on the positive	Nurture yourself	Laugh at yourself
Have more fun	Ask for help when you need it	Address cognitive distortions
Have a good cry	Spend time in nature	Address conflicts with others
Connect with others	Manage time better	Yoga

45   Joseph Mercola DO website Nov 07 2014 fitness.mercola.com

Stress Reducing Meditation

Meditate asking the Santa Maria to open your heart.

Let yourself feel whatever is coming up.

Just allow it to be whatever it is with no judgment.

Take as long as you need.

Say out loud:

When I wallow in self pity -

I feel like a victim with no way out.

When I wallow in self-hatred

I become my own abuser.

When I allow myself to accept self compassion - I feel loved

and whole

With Hope that this too shall pass.

When you are ready

With pursed lips blow off stressful feelings. [46]

Think of someone or something you love with all your heart

Now extend that love to yourself

Allow yourself to be surrounded by love and compassion

For where you are right now.

Let it open your heart with warmth.

---

46   This is the breath work I discussed earlier where we pretend we have a
drinking straw in our mouth and we breathe through it only. This slows down our
breath and creates a "back draft" to our lungs.

11 Cannabinoids Especially Cannabidiol (CBD)

Cannabinoid basics part 1

In this chapter I address one of the two principle chemical types in cannabis: cannabidiol (CBD) and introduce the Endocannabinoid System within which MM works. This is *Cannabinoid Basics* part 1.

Then in Appendices 6 and 12 I will describe THC, how it acts and interacts with CBDs within the brain and body (pages 157 and 181-2) as Cannabinoid Basics part 2 and part 3.

"Medicinal Marijuana" is remarkably different than the commonly prescribed single "drug" that our physician may prescribe for us at an office or hospital visit. Those unnatural drugs are synthesized and made by chemists from other chemicals. By contrast, MM uses the natural cannabis plant and its two basic cannabinoids: tetrahydrocannabinol (THC) and CBD as medical therapy to lessen symptoms, treat problems and possibly help lessen the physical and emotional pain of some diseases. Using MM requires taking more personal responsibility in appropriate aspects of our healing process.

Endocannabinoid system *Part 1* In the early 1990s this was found to be one of the body's most important physiological and neurotransmitter systems involved in establishing and maintaining our natural physiology and balance, also called homeostasis. Undiscovered until 1988-92, scientists have now shown that this group of natural chemicals and their receptors are found throughout the human body and are instrumental in a variety of physiological processes including pain modulation, memory, appetite and more.

Because knowing about CBDs and the Endocannabinoid System (**ECS**) will be helpful to understanding how to use MM, I will address these first and later describe the role of THC. This chapter may be more technical and may be difficult to understand. I will describe chemicals in the brain that MM helps to balance, according to the latest research. Skip it if you are not interested or look at the documentary described in the next paragraph. (For those that are interested, I have put the most technical explanations for cannabin**oids** and cannabi**diol** in Appendices 6 and 12.)

Here is the link to an easily understandable 1 hour 10 minute documentary published in April of 2014. Some of the main researchers who are working diligently to prove much of what is listed in the pie chart at the end of this chapter are in this film: Marijuana Documentary April 20, 2014–Cannabis Research Studies youtu.be/sYA9EpVB2qo.

MM writer Martin Lee published 'New Light on the Darkness of PTSD' on MAPS.org in late 2013 and his website Project CBD: "A recent article in the journal *Neuroendocrinology* highlights the crucial role of the endocannabinoid system in protecting against posttraumatic stress disorder (PTSD), a debilitating chronic condition involving horrific memories that cannot be erased. In an effort to understand the neurobiological mechanisms that underlie the onset and development of PTSD, a team of U.S. and Canadian scientists analyzed 46 subjects who were near the World Trade Center in New York City during the September 11 terrorist attacks. Twenty-four of these subjects suffered from PTSD following the attacks; 22 subjects did not.

The researchers found that people with PTSD had lower serum levels of anandamide (also called AEA), an endogenous cannabinoid, compared to those who did not show signs of PTSD after 9/11. Innate to all mammals, anandamide (our inner cannabis, so to speak) triggers the same brain receptors that are activated by THC and other components of the marijuana plant." [47]

This article concludes in Appendix 7 on page 170.

Cannabinoids are a class of diverse chemical compounds that act on cannabinoid receptors in our brain and most other organs. These receptors are analogous to the lock and cannabinoids are analogous to the key that 'opens' the lock and brings about the effects on the brain and body.

Receptor / Lock   **Cannabinoid / Key**

See an illustration of the endocannabinoid system (ECS) on the next page.

---

47   MAPS Annual Report maps.org/news-letters/v23n3/v23n3_40-42.pdf and projectcbd.org/news/cannabinoid-sci-ence-sheds-new-light-on-the- darkness-of-ptsd/

112

The Human Endocannabinoid System

CBD, CBN and THC fit like a lock and key into existing human receptors. These receptors are part of the endocannabinoid system which impact physiological processes affecting pain modulation, memory, and appetite plus anti-inflammatory effects and other immune system responses. The endocannabinoid system comprises two types of receptors, CB1 and CB2, which serve distinct functions in human health and well-being.

Receptors are found on cell surfaces

CB1 receptors are primarily found in the brain and central nervous system, and to a lesser extent in other tissues.

CB1

THC — Tetrahydrocannabinol

CBD does not directly "fit" CB1 or CB2 receptors but has powerful indirect effects still being studied.

CBD — Cannabidiol

CB2

CBN — Cannabinol

CB2 receptors are mostly in the peripheral organs especially cells associated with the immune system.

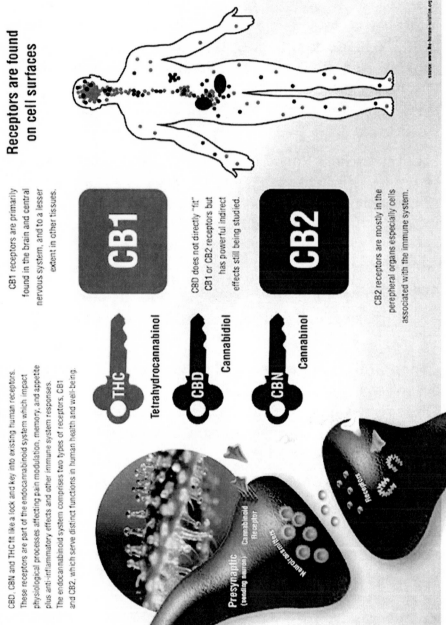

Presynaptic (sending neuron)

Cannabinoid Receptor

Neurotransmitters

Depending on the molecule, this action then 'opens' or 'closes' the lock. When the lock is open a physiologic reaction occurs to activate the body's neurotransmitters. These may be excitatory, such as dopamine (an 'upper' or stimulant) or inhibitory, such as serotonin (a 'downer' or sedative).

At least 85 different cannabinoids have been isolated from cannabis and up to some 200 flavenoids/terpenes that appear to influence its effect. Over the past 30 years researchers have evaluated many of these. We await their eventual discovery of how each individually or in combination can be most effective in treating various health problems.

In PTSD, the part of the brain called the amygdala malfunctions to trigger anxiety. The amygdala is one of six parts of the brain involved in cannabinoid effects involving the CB1 receptor. [48]

Instead of being able to block memories of highly traumatic events, it malfunctions and lets the person be flooded with terrifying emotionally painful memories. The affected person repeatedly remembers and relives the terror. (But knowing these facts can help us heal around their related pain. We do so by naming our traumas and then sharing our experiences and pain with safe others.)

---

48   Distribution of CB1 Receptors in the brain: 1) *Hippocampus* –Memory and Learning, 2) *Amygdala* – Novelty, Emotion, Appetitive Behavior, 3) *Basal Ganglia* and Motor *Cerebellum* –
Real Time Coordination, Selective Attention and Time Sense,
4) *Nucleus Accumbens* - Reward Mechanisms, 5) *Cortex* and *Frontal Lobe* - Executive Function, Judgment, Synthesis, and Evaluation - from Center for Medicinal Cannabis Research (CMCR) U Cal, S Diego

This malfunction can be attributed to a faulty cannabinoid receptor mechanism and process.

Scientists have suggested two possible causes:

1) The amygdala's cannabinoid receptors don't take in the body's natural cannabinoids because they are either one or more of

a) **malfunctioning** or b) **limited** in **number** and/or **efficiency**. Or c) both.

2) The brain and the body cannot make enough of its natural cannabinoids to quiet this repeated terror response. Either of these situations results in the continued state of hyper-arousal with the resulting fight-or-flight feeling or "walking on egg shells." Any illness of fear or terror may be helped by CBD. This includes schizophrenia and some other psychoses which Bertram Karan PhD calls an illness of terror.

Endocannabinoid System *Part 2* Endo means that it is produced naturally in the human body. [49] This system consists of a group of neuro-modulatory (effecting neurons) lipids and their receptors in the brain and other organs that are involved in a variety of physiological processes including appetite, pain-sensation, mood, and memory.

From the 1950s only a few research laboratories had studied marijuana. Then the Israeli pioneer Raphael Mechoulam PhD and colleagues discovered the body's inborn endocannabinoid system (ECS) and one 'THC like' molecule that they named Anandamide. [50] They found that this new messenger molecule played a role in pain.

---

49   "Endo" also means within the human body's system of cannabinoid receptors.
50   Named by Dr. Mechoulam and his colleague Roger Pertwee in 1992 as anandamide based on the Sanskrit word ananda which means supreme joy or bliss for
its feel good quality. William Devane and Lumír Ondřej Hanuš inspired the name.

Since then thousands of researchers worldwide have studied the molecular physiology of cannabinoid receptors in laboratory animals. This research is focused less on cannabis and more on the nature of the ECS within the animal and human body. They have found these CB specific receptors on the cell membranes of most cells in most organs of our body and in animals. More: they have found that this system, especially the CB1, may be the most physiologically active and important system in the body. If we have the receptors, then we must also make the chemical in our own body or we wouldn't have receptors for them. [51] We have more to learn.

Our having the Endocannabinoid System inborn in our body is even more evidence that cannabis/cannabinoids are a natural medicine, and that our body is set up to take in and possibly help balance us physically, mentally emotionally and spiritually in varying degrees.

Cannabis has far more of these cannabinoids in its leaves, seeds, buds and stems than we do in our bodies. There are intriguing possibilities for cannabinoid based medicines to regulate the balance (homeostasis) of our body.

Cannabinoids are from a natural botanical plant [52] and have potentially healing physiological actions. As I discuss throughout this book, CBD oil taken in drops or vaped can lessen anxiety without making us "high" or disabled. [53]

---

51   Mechoulam R, Hanus L Anandamide and more. ed. Holland J *The Pot Book*: Its Role in Medicine, Politics, Science, and Culture. ParkStreetPress, VT p. 71

52   Examples of other drugs that are found in nature and work in our bodies include: opium, aspirin, digitalis, and the cancer chemotherapy vinca alkaloids of vincristin and vinblastin.

53   CBD products may contain trace amounts of THC but not enough to produce a "High."

Our bodies are hardwired from birth to accept and use CBD for our good. Scientific research is proving that this system creates balance not only in our brain but also in most of our organs. The cannabinoid system may be a useful part of our present and future for helping to balance our mind and body. Understanding it and using cannabinoids carefully and appropriately may also help our planet by not adding toxins to our environment. [54]

Cannabidiol (CBD)

Cannabidiol in oil or paste is a blend of CBD, CBDA, CBC, and CBG, so we can call it by the singular CBD or pleural CBDs if we prefer. It counteracts some of THC's effects, including paranoia and anxiety. A major way it may work is that CBD inhibits the enzyme FAAH (fatty acid amide hydrolase) which breaks down our body's natural anti-anxiety neurotransmitter anandamide. CBD also inhibits glutamate toxicity in hypoxic (low oxygen) states such as in stroke and other injuries. Some drug companies are quietly looking for a new drug that also does that. It may be difficult to find a better healing and protective agent than the natural CBD from this plant.

CBD has shown promise in the easing of symptoms of a wide variety of conditions, including PTSD, insomnia, opiate and other drug withdrawal, rheumatoid arthritis, diabetes, autism spectrum disorders and epilepsy.

I elaborate on a few of these below.

---

54   There is even consideration here for the planet because we are finding that many of the toxic chemicals manufactured by the pharmaceutical industry pollute our water supply. In the US there is much debate that in some places the water has anti-depressant and other drugs in it. There is also fear that antibiotics have not only polluted our waters, but our bodies have become resistant to their help.

CBD is currently being studied for its anti-cancer potential. (See, e.g., pie chart based on scientific research on page 125.) CBD remains a legal dietary supplement and the advice on some bottles may say: "Use sparingly: a little goes a long way." Its half life smoked is 1-1.5 days and when ingested 2-5 days. Taking more CBD than needed may result in over sedation and/or an unpleasant or queasy feeling. Cutting back the dose should stop that.

People using MM for chronic pain or anxiety may be best to consider a form with a ratio that is less psychoactive and more physical in effect, such as products with higher CBD content vs. THC. (More in Appendices 6, 12, and 13).

CBD also works well to settle someone down if they've had too much MM with THC and are uncomfortable. One of the reasons the hemp oil works well is because it has little or no THC. Use less than the label recommends and if it doesn't work within 10-20 minutes, use a little more. Do not exceed the recommended dose. This will balance out the THC/CBD ratio and many to most people will have a positive response.

In 1963 Mechoulam isolated cannabidiol which is now known to be one of at least 85 active cannabinoids identified in cannabis. He reported that CBD caused no marijuana-type "high" effects but that it caused other important effects. For example, it is a potent anti-inflammatory drug, it arrests the onset of autoimmune diabetes, ameliorates cognitive and motor impairments in animal studies via activation of 5-HT1A receptors and it reduces epileptic attacks and lowers anxiety. To understand CBD better, see bottom of page 192 for a recent Dateline NBC special on CBD (and in one case, THC-A) helping children with intractable seizures, which you can watch online.

CBD for PTSD, Sleep and Drug Withdrawal

Many **PTSD** patients who have used the glycerol based Hemp Oil (that contains at least 400 to 500 mg CBD) report a new sense of calm and peace, which they didn't expect. It can be taken on or under the tongue or inhaled in a similar inexpensive vaporizer as nicotine dependent people use to stop smoking cigarettes.

Others and I have seen many opiate, benzo and other drug dependent patients use CBD in a similar way to help them taper off the drugs. Used the above way CBD commonly lessens and may eliminate the **drug withdrawal** symptoms for several hours, after which another dose works again to relieve the bothersome drug withdrawal. One cannot usually 'overdose' on CBD.

Derived from the hemp strain of the cannabis plant stalk, stem and seeds, hemp oil with CBD is sold as a food nutrient or dietary supplement, delivering the benefits of cannabinoids without the psychoactive/euphoric effects of THC. It has been legal in all U.S. states since 2010.

Cannabis and hemp contain over 400 different molecules, with the 85 active cannabinoids mentioned above and some 200 or more flavonoid/terpenoids and other yet identified molecules. Different strains have different amounts and ratios of these. The present 'dietary supplement' products contain hemp whole plant extract from isolated strains or 'cultivars' of hemp that provide significant quantities and concentrations of CBD. Some PTSD patients taking CBD report tapering down to the least amount their body needs to stay calm, usually taken twice a day. If they have gone beyond the 6 to 12 hours, their body may remind them to take it, with their prior PTSD or drug withdrawal symptoms returning and now possibly benefiting from another dose.

Problems sleeping are common in both PTSD and opiate and other drug withdrawal. These can be helped by using the above CBD schedule including a dose at bedtime. To supplement that I have seen physicians use other safe sleep supplements (as in the book *Not Crazy* by Whitfield) and clonidine 0.1 to 0.3 mg. Melatonin in oil may be helpful if used in a separate simple inexpensive GE 4 'tank' or similar vaporizing device.

For THC-induced panic attacks: CBD alone in oil is useful to have in case the patient that is new to MM panics or is 'too high.' CBD drops under the tongue for 60–90 seconds or vaped will usually bring them back to a normal state because the CBD balances a THC excess reaction. The CBD drops will thus usually shift the percentage or balance of THC-to-CBD toward or to a normal waking consciousness.

Six pages below is a potentially useful **Pie Chart** of various cannabinoids and what researchers are studying and have found. It is commonly shown on the Internet. I tried to find an original citation for it and couldn't. So far there aren't enough studies for scientific replication to convincingly validate all of these claims. For a credible explanation that may back the Pie Chart see Appendix 12 Tables A 9 and 10 that was based in some part on 74 peer-reviewed published scientific and clinical studies compiled by Cannlabs of Denver in 2012. In spite of the Federal government and NIDA's limiting cannabis research since 1940, scientists worldwide continue basic molecular, biologic and some clinical research addressing MM.

Cannabis Dosing Guidelines

Using Medicinal Marijuana is *personalized* medicine.

It is personalized because the appropriate dose depends on the person, problem and condition being treated.

Emphasizing what I say throughout this book, MM clinical observers Fred Gardner and Martin Lee suggest the dosing guidelines in the box below. [55]

"CBD and THC share a special interdependent relationship and work together to increase one another's therapeutic benefits. CBD is a non-psychoactive compound. THC is psychoactive and, therefore, may produce euphoric or dysphoric effects. A patient's sensitivity to THC is a key factor in determining appropriate dosages and ratios for a CBD-rich treatment regimen. CBD can lessen or neutralize the psychoactivity of THC. So an increased ratio of CBD-to-THC means fewer mental effects.

1. Decide how you want to take cannabis. Dosed medicine infused with concentrated cannabis oil extracts is available in sublingual sprays, capsules, edibles, tinctures, and other products.

2. Find your ratio. Cannabis products have varying amounts of CBD and THC. A high CBD strain or product (with little THC) is not necessarily superior to one with a more balanced CBD:THC ratio. Find the proper combination for you.

3. Begin with a low dose—especially if you have little or no experience with cannabis.

---

55   From Gardner and Lee accessed 10 Nov 2014 projectcbd.org/ medicine-2/ dosage/

4.Take a few small doses over the course of the day rather than one big dose.

5.Use the same dose and ratio for several days.

Observe the effects and consider if you need to adjust the ratio or amount.

6.Don't overdo it. Often with cannabinoid therapeutics, "less is more." Cannabinoid compounds have biphasic properties. This means that higher doses can sometimes be less effective than low or moderate ones. Also, too much THC—while not lethal—can increase anxiety and mood disorders.

7.Consider the condition you're treating. For anxiety, depression, spasms, and pediatric seizure disorders, you may do better with a moderate dose of a CBD-dominant remedy. Look for a CBD:THC ratio of more than 14:1. For cancer or pain, you may need more THC, for instance, a 1:1 ratio." [56]

Individual Sensitivity

A patient's individual sensitivity to THC is a key factor in determining appropriate dosages for a CBD-rich treatment regimen. MM and cannabis therapeutics is personalized medicine. The appropriate dose depends upon the person, the problem or condition being treated and their individual response to using MM.

Our best choice then may be to use a personal trial and error method as we ingest MM each time, while keeping an accurate

56   From Gardner and Lee, accessed 10 Nov 2014 projectcbd.org/ medicine-2/ dosage/

diary or log of MM type, delivery method, dose and result for each problem for which we are using it. Doing so will help you and others (if you ever share these details) to have a better understanding of what worked for you and what did not. See Appendix 2 for a full page of this table below to copy for future use.

Table 5 My Personal MM Use – A Trial and Error Results Log

Date	Problem	MM Type	Method	Dose	Results
10 Oct	Chronic fear & anxiety*	CBD drops	Deposit under tongue	5 to 15 drops 1-2 x/day**	Quick calm
15 Nov	Opiate withdraw-al (WD)	CBD oil/ drops	Vaped q 4-8 h prn	One in-hale	Less WD symp-toms
16 Dec	Can't sleep	CBD oil/ drops	Vape bed time	One in-hale	Sleep

*Assuming otherwise treatable causes are ruled out.
**See instructions in oil or drops literature.

Patterns in Research and Clinical Practice

When we study the archives of scientific research, medicine and epidemiology we find a repeating pattern of *discovery*, *trials* of one or more new treatments, our eventual realization about the sequence of events, and then more research and trials. But there is one common, if not universal pattern wherein we observe and realize the reality of our theories and ultimately our clinical practice.

This repeating pattern has been summarized over the centuries including our most recent decades from 1960 until today in various ways. To me the clearest and simplest pattern is that we can divide these into three phases of research, development and practice:

1) The *Gee Whiz* phase (Cannabis, as MM, will cure our every ailment)

2) The *Aw Nuts* phase (Cannabis, as MM, doesn't work on every ailment to perfection in every person)

3) The *Yes-But* phase (This is where most of us are today for most treatments in most of the practice of medicine.)

Skeptics have usually helped us look at both sides. (See in the references skeptics on cannabis who bring up many websites that show how little we still know about this plant, in large part because of the politics and laws against our researching its uses). I know from my own experiences, clinical practice and what others have reported about their illness and experiences, that we are about to legalize a plant/herb/botanical and its various chemicals, that appears to be an important advance that will likely change some of the ways our medical model has worked and not worked for many decades. MM is moving us further into the paradigm of Holistic Health which The Recovery Movement has been using since the mid-1980s. We know that this stage-oriented approach works in individual and group clinical practice.

Cannabis as MM helps us experience a more expanded level of consciousness. When we apply MM as others and I describe, doing so can give us the tools to take responsibility for us and our patients to maintain our health instead of using medical doctors alone. Knowledgeable physicians, DOs, NDs, and DCs can advise us when we find the one(s) who know(s) MM.

CBD Dosing: In the next table I list examples of CBD doses in mg per milliliter of oil that I have found to be potentially useful.

CBD Dosing in mg per ml - *Examples only*

mg/ml	*Dose Obtained*	*Source example*	*Comment*
3 mg	30 mg/10 ml	Miraclesmoke.com	Discounts in some. Check online for current cost/mg Patient still has to use trial and error process
4.5	45 mg/10 ml		
8.3	500 mg/60 ml	Generic brands, e.g., Cibdex, Ultra, Dixie, Bluebird, Tasty	
10 20	250 mg/25 ml 60 mg/3 ml	Hemplifetoday.com	
50 to 150	50 to 150 mg/10 ml	Endoca hemp oil makes next highest mg/ml I have found	May help if above mg/ml doesn't. Some find the 656mg/ml oil to be the strongest.
656	39.4 gm/60 ml	Cannabis oil by Regalabs	

I started with the 8.3 mg/ml dose and found that the higher mg doses worked better. If a lower CBD dose is not effective for a symptom or problem, some patients may want to try an oil with a higher dose. The 656 mg/ml oil has an olive oil base that may vape better if thinned with a few drops of glycerol.

Learning by trial and error is similar to how we use most conventional single molecule drugs. Since CBD is from the hemp plant it usually has several other cannabinoids that the makers' websites may list and some discuss.

Adding THC complicates the process of using MM and requires more of the patient's attention to plan and try the herb or its extract, as discussed in App 13 p. 187. The Pie chart on the next page gives a representation of how helpful CBD can be.

**Figure 3 Pie Chart of Cannabinoids
and Claimed Usefulness**
See page 182 *and following* for discussion and summary of
documentation of many of these findings.

The pie chart is easily found on the Internet. I looked for its origin and couldn't find it. Then I looked at the research on its claims and was pleasantly surprised to find enough completed and still-in-progress projects on each of these items. (The documentary at the beginning of this chapter contains many of these researchers and explanations of their research.)

The chart summarizes the healing effects found to date for each cannabis component. CBD is clearly the most effective. At times it needs to be in a proper ratio with the others to help heal other illness and problems. Clinical scientists around the world are studying these observations. One example among many are at the Center for Medicinal Cannabis Research, University of California San Diego (cmcr.ucsd.edu)

Possible CBD Use for *Relapse Prevention* and/or *Drug Detoxification* and *Dose Tapering*

Mechoulam and Parker noted "In the cell, anandamide is hydrolyzed [broken down] to arachidonic acid and ethanolamine by fatty acid amide hydrolase (FAAH).

2-AG is also hydrolyzed enzymatically, both by FAAH and by mono-acyl hydrolases. Suppression of these enzymes prolongs the activity of the endocannabinoids" [which researchers and many in Big Pharma are studying].

In their section on Cannabinoids and Relapse they say (my summary): A growing literature suggests that Inhibiting FAAH (the above enzyme that breaks down anandamide) and by giving CBD by mouth or vaporizer also relieves opiate drug withdrawal and prevents relapse due to drug craving, seeking and using, detailed below.

"Cannabidiol, also attenuated cue-induced heroin seeking as well as restored disturbances of glutamatergic and endocannabinoid systems in the accumbens produced by heroin seeking. Apparently, in addition to the many other ailments that cannabidiol improves it may also be a potential treatment for heroin craving and relapse."

Whether these observations bear fruit will depend on how other researchers, clinicians and drug dependent people experiment by using their various forms of trial and error, a basic principle in most of MM that I describe throughout this book.

Mechoulam and Parker The endocannabinoid system and the brain. *Annual Review of Psychology* 64: 21-47 January 2013

12 Spiritually Transformative Experiences (STEs)

H ave you had a spiritual awakening? Or do you wonder if you might have had one?

A spiritual awakening is an experiential opening to a Power greater than ourselves.

Triggers for Spiritually Transformative Experiences

Based on my and my colleagues' research [57] and informal surveys that I have done with people attending my talks and workshops over the years, I estimate that at least one in three people have had a spiritual awakening of some sort. Perhaps 25% of those spiritual awakenings were triggered by near-death experiences.

The remaining 75 % are triggered by numerous other experiences, from meditation, to childbirth, to ''hitting bottom'' in a critical or desperate life situation. Some of these events have opened some people to experiencing

---

57   Bruce Greyson MD, Kenneth Ring PhD, Charles Flynn PhD, Charles Whitfield MD, Lawrence Edwards PhD

the sometime painful yet often freeing spiritually

transformative process. [58]

Being near-death is one trigger for a spiritual experience.

Other triggers include:

Withdrawing from alcohol or drugs (detox)

Childbirth

Experiencing the Death of a loved one

Great loss

Reading spiritual literature or hearing a spiritual talk

Intense prayer or meditation

An encounter with Angels or other beings

An intense transcendent sexual experience

Spontaneously when in nature

In a big dream that is remembered for life

Kundalini Awakening

Breath and body work [59]

58, 59   See *Spiritual Awakenings*: Insights of the Near-Death Experience and Other Doorways to Our Soul. 1995 Health Communications, Deerfield Beach, FL

Taking a Psychedelic (mind expanding) drug

During an intense trauma

Gradually without first having a dramatic experience

And by

Using Medicinal Marijuana over time when chronically ill or when pursuing personal growth and/or Stage Two Recovery as shown in chapter 6 on page 55-6.

As a result from having experienced any of these, we become more aware of and open to our self, others and the Divine Mystery.

Who or what is it that actually does the awakening? Is there a part of us that begins to become more aware and opens to our self, others and the God of our understanding? [60]

My sense is that it is a spiritual energy that I and others believe is from God that starts to awaken us to our Real or True Self, and helps us learn about how our ego/false self blocks that. [60]

Who Am I? A Map of the Sacred Person

Throughout the struggle of the human condition, many people have asked some important questions: Who am I? What am I doing here? Where am I going? How can I get any peace? While the answers to these questions remain a Divine Mystery, we have found it useful to construct a map of the mind or psyche, summarized on the next page.

---

60 Harris Whitfield B 2009 *The Natural Soul*. Muse House Press, Atlanta

Figure 4

Map of the Sacred Person

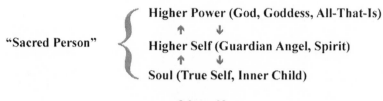

Excerpted from *The Natural Soul*
Copyright © 2009, Barbara Harris

And while the map is not the territory, maps can be useful.

Other names for or dimensions of the True Self, who we really are, include the real or existential self, the human heart, the Soul, chakras 4 and 5, and the Child Within. They are all the same because they are our True Identity.

We each also have within us a Divine Nature, sometimes called "Holy Spirit", guardian angel, Ruach haKodesh, [61] Atman, Buddha Nature, Christ Consciousness, Higher Self, or simply Self. And both of these, our True Self and Higher Self are intimately connected to Higher Power, God/God- dess/All-That-is, a part of which is also within each of us.

_____

61   Ruach haKodesh is the closest Hebrew term for "The Holy Spirit" in Christianity. It is used as our own personal "Higher Self" or "Higher Power." However It is used, It connects us Up (with a capital U!)

We see this relationship---True Self, Higher Self and Higher Power as being such an important relationship that we can also view it as being one person, which we call the Sacred Person. In a loving, supporting and teaching way, pervading throughout the Sacred Person, is God's Holy Spirit (also called Kundalini, Chi, Ki, Ruach haKadosh or Divine Energy).

As a part of the Mystery, our True Self makes, constructs or allows an assistant to help us survive and help us in other limited ways. We can call this assistant, this side-kick, the ego also known as the false self or co-dependent self. When this ego is helpful to us, such as in screening, sorting and handling many aspects of our internal and external reality, we can call it positive ego. But when it tries to take over and run our life, it becomes negative ego. [62]

This map of the psyche is more evolved than the maps of Freud, Jung and their colleagues of up to 100 years ago, when they used the term "ego" to mean both True Self and false self. Since the 1930s the self psychologists and the object relations psychologists have begun to make this more precise differentiation between True Self and false self and today we use '"ego" synonymously with false self.

To clarify: This updated understanding is in contrast with many writers who still lump the True Self and false self together and erroneously call both the "ego."

---

62 See Appendix 9 for traits of the ego vs Traits of the Soul. We developed ego to assist our True Self or Soul. Our ego is the part of us that is afraid to die. Our True Self/Soul is divine and infinite. Therefore when we learn to let go of our ego's beliefs and live as our Real Self/Soul, we can no longer fear death.

A contemporary holy book called *A Course in Miracles* says in its simple introduction:

What is real cannot be threatened.

What is unreal does not exist.

Herein lies the peace of God.

What is real is God and God's world, that of the Sacred Person. The ego and its world are not real, and therefore, in the grand scheme of the Mystery, does not exist. Herein, when we make this differentiation, lies our peace and serenity.

But growing up in a dysfunctional/unhealthy family and society of origin, we may have become traumatized and wounded. That wounding made our True Self/Soul go into hiding, and the only one left to run the show of our life was our ego (false self). And since it is not able or competent to run our life successfully, we often end up feeling confused and hurt.

The way out is to begin to differentiate between identifying with my True Self and my false self, and to heal my wounds around all of the past traumas that hurt and confused me. That is what I have described in my other books, especially Spiritual Awakenings and The Natural Soul.

While all of this information is useful to know on a cognitive level, it is healing only on an experiential level. To heal, we have to experience working through our pain, as well as living and enjoying our life anew. If we can identify with having a spiritually transformative "arousal," that gives us the gift of co-operation by a spiritual energy that can assist us as we heal.

Spiritual Awakenings and
the Recovery Movement

Over the decades of the 1980s and l990s, and into the 21st century, an increasing number of people have begun to awaken to many of their traumatic experiences and are beginning to heal themselves. This phenomenon, called the recovery movement, with its free and effective Twelve Step Fellowships, is part of a new paradigm, a new and expanded understanding and belief about the human condition and how to heal it. This approach is so effective and has developed so much momentum for four reasons: 1) it is grass roots - its energy comes from the recovering people themselves, 2) it is based on stages of recovery (page 54), and 3) it employs the most accurate and healing of all the accumulated knowledge about the human condition. [63] But what is different about this knowledge is that

4) it is simplified and demystified.

Traps in Spiritual Awakenings

When we have a spiritual awakening, we have an expansion of our consciousness. We see and experience our inner life and outer life and world in a more expansive way. While most of our raised consciousness will be positive, there are some traps at times, especially if we tell people who don't understand. After having had a spiritual awakening and a possible associated energy arousal, one trap is 1) being misled by other people

---

[63]   See References for *Healing the Child Within, A Gift to Myself* and *Wisdom to Know the Difference*: Core Issues by C Whitfied MD

who may try to invalidate our experience and steer us off of our personal spiritual path. [64]

When we allow our healing process to evolve naturally when we use a full recovery program, the result is usually continued psychological and spiritual growth over time. A problem is that many of the symptoms and experiences mimic what psychiatry and psychology calls "mental disorders." Most psychiatrists and some psychologists, social workers and counselors aren't able to recognize these breakthrough experiences[65] and instead they prescribe or recommend one or more psychiatric drugs in an attempt to lessen the patient's symptoms. By contrast, in these awakenings, we become progressively more connected with self, others and God. Most psychiatric drugs are toxic to the brain and body and tend to shut down or aggravate the normal flow of this process. The drugs slam the door shut to our psycho-spiritual growth.

Psychiatrist Peter Breggin, MD said, "It is difficult, if not impossible, to determine accurately the psychological condition of a person who is taking psychiatric drugs. There are too many complicating factors, including the drug's brain-disabling effect, the brain's compensatory reactions and the patient's psychological responses to taking the drug. I have evaluated many cases in which patients have deteriorated under the onslaught of multiple psychiatric drugs without the prescribing physicians attributing the patient's decline to drug toxicity.

---

64   See References for *Spiritual Awakenings*: Insights of the Near-Death Experience and Other Doorways to Our Soul by B Harris Whitfied
65   There is a group that educates mental health professionals on helping patients integrate spiritual experiences. They have a list of trained professionals all over the US. It is the American Center for the Integration of Spiritually Transformative Experiences (ACISTE). Website is aciste.org

Instead, physicians typically attribute their patients' worsening condition to 'mental illness' when in reality the patient is suffering from adverse drug reactions. [66]

Another trap is the frustration that usually comes with trying to do what is called a spiritual bypass. A spiritual bypass happens when we try to avoid working through the pain of our prior traumas, (that could also be holding down our immune system and allowing us to get one illness after another) so that we may try to jump from an earlier stage of healing directly into the most advanced stage. Because this concept is crucial to making sense of and handling spiritual awakenings and the movement of spiritual energy you may want to go back to page 53 and read the stages of recovery again.

Spiritual Bypass (also called "Premature Transcendence" or "High Level Denial")

If we try to go around or bypass the darkness to get to the Light, i.e., if we try to ignore the lower to get to the higher levels of our consciousness, something—we can call it our shadow (Jung) or repetition compulsion (Freud)—will pull us back until we work through our particular unfinished business. Trying to avoid this work of Stages One and Two recovery can also be called premature transcendence or high-level denial. This is seen in any number of situations, from being prematurely 'born again,' to having a spiritual awakening and focusing only on the Light, to becoming attached to one way that is the "only" way.

---

66   I advise against starting psychiatric drugs. If you are already on them, do not stop suddenly. You need to be guided by your health care professional to taper slowly. sometimes taking months.
(See C Whitfield's book *Not Crazy*: You May Not Be Mentally Ill.)

The consequences of spiritual bypass are often active co-dependence: denial of the richness of our inner life; trying to control one's self or others; all-or-none thinking and behaving; feelings of fear, shame and confusion; high tolerance of inappropriate behavior; frustration, addiction, compulsion, relapse, and unnecessary suffering.

The way out of this trap is to develop humility (i.e., openness and willingness to learning more about self, others and God) and work through the pain of wherever we may be, or just enjoy the joyous feelings. Those who are actively addicted or disordered can work through a Stage One full recovery program. Those who are adult children of troubled or dysfunctional families can work through Stage Two recovery. We need to stay mindful of these necessities: we cannot let go of something if we do not know experientially what it is that we are letting go.

We cannot transcend the unhealed; and we cannot connect experientially to the God of our understanding until we know our True Self/Soul, our human Heart. People who have progressed in their STE Process may want to join a Stage Two or Three therapy group to help support themselves in their new experience of co-creating their life with a Higher Power. Their new expansive and creative abilities may not fit in to their original life/relationships and being in a weekly therapy group gives them a place to talk about their feelings and check themselves out with fellow aspirants.

Integration and Liberation

Stage Three recovery is the one into which we may be compelled prematurely by having a spiritual awakening. If we haven't previously done our Stage Two work to heal any of our trauma experiences' effects and issues, it will be hard to live as easily on a Stage Three level of consciousness.

When we have worked through our unfinished business, our everyday experience includes a healthy balance and natural spirituality. This is an ongoing process. In this stage we make meaning out of our past and bring that with positive insight into our present. We are now aware of being free of old beliefs and can use this stage to get comfortable with a fresh outlook that is creating stability in our life while practicing gratitude and humility.

If we can expand our beliefs and bring our higher nature into our everyday life—we can begin to experience true humility. Growing in this way invites us to stretch beyond the limits of who we thought we were and become all that we are. This process allows us to experience a healing unity within ourselves, others and our Higher Power. It brings us to the point where we are not living in our heads anymore. We are living in the present moment and conscious of our mind, body and spirit.

Our intuition ripens so that we have *another way* of *knowing our reality*. Our natural but well thought out intuition coupled with this plant we call Medicinal Marijuana or Santa Maria has the healing properties to *show us another way of knowing* our *inner* and *outer reality.*

Three pioneer psychedelic explorers said "Of course, the *drug* does not produce the transcendent experience. It merely acts as a chemical key. It opens the mind and frees the nervous system of its ordinary patterns and structures. [67]

"Psychoactive drugs can only interact with the substrate that is already present in your brain when you take the drug. The nature of the experience depends almost entirely on set and setting. Set denotes the preparation of the individual, including his personality structure and his mood at the time. Setting is *physical* - the weather, the room's atmosphere; *social* - feelings of persons present towards one another; and *cultural* - prevailing views as to what is real. It is for this reason that manuals or guide books are necessary. Their purpose is to enable a person to understand the new realities of the expanded consciousness and to serve as a road map ." [67]

These wise words are in part what using cannabis is about, from using it medicinally to recreationally. Some have called it a kind of Swiss Army knife where it can have various effects on various human problems. But all this is coming *from within us*, not directly from the drug. It is already and always an awakening and an ongoing process of our inherent divinity in our authentic and sacred relationship with our real self, the real self of others and the God of our understanding. All of this makes me wonder if somewhere in our evolution there was the set up for the endocannabinoid system to eventually help us wake up, to expand and open up spiritually -- as well as to become balanced physically, mentally and emotionally.

*   *   *

---

67   Leary T, Metzner R, Alpert R 1964 *The Psychedelic Experience.* University Books. New York, NY

**Closeup photo of a cannabis flower
which is coated with trichomes containing cannabinoids**

Trichomes are hair-like extentions that are rich in the cannabinoids that I describe and discuss throughout this book. Countless recearch scientists today continue to study their healng properties and even more countless amateur 'scientist' users continue to employ it to help them explore their consciousness, health and lives. An additional number are now using it as MM and some are exploring how to make that work better to help them restore several important aspects of their health.

Why do most Republican and Democrat politicians fear marijuana legalization? Well, while they do line their pockets with money from industries that benefit from prohibition, there is an underlying truth to marijuana consumption.

When someone starts smoking marijuana, they realize that the entire political system is rigged against them. They realize that "Left" or "Right" they are still merely hands of the "same beast".

In turn, cannabis smokers begin a process of breaking down their "left-right" paradigms and begin to advocate civil liberties, smaller more localized government and advocate "Freedom" above concepts such as "Security".

As you can imagine, this scares the living hell out of those "fat cats" that benefit from prohibition because as more people migrate away from the policies that keep them in power, their careers are on the road to extinction.

Cannabis does change your mind. No matter how you look at it, most people who smoke marijuana become more tolerant of others, they lose the paradigms that divide humanity, they implore critical thinking and become highly skeptical of mainstream news and government.

So you see...marijuana is INCREDIBLY DANGEROUS, for politicians, pharmaceutical companies, timber, oil, plastics, alcohol, tobacco, cartels, banks and so on.

These are a handful of people who dominate the world we live in. They are the 'kings of old' who would rape and pillage because it's their "divine right" or so they believe.

So every time you spark one up, know that you are committing a heroic act of defiance to a system that depends on enslaving people who exercise their freedom of consumption!

from MARIJUANADoctors.com

Epilogue

"If the words, 'life, liberty and the pursuit of happiness' don't include the right to experiment with your own consciousness, then the Declaration of Independence isn't worth the hemp it was written on."

Terence McKenna

It may be ironic, even comical, that I am working in the addiction and recovery movement and I am writing a book on medicinal marijuana. However, it's not me that has changed. It's our laws and our evolving understanding of the reality and potential of this plant that are changing. I could have written this book ten years ago if we were as advanced as we are now under the umbrella of "Medicinal" Marijuana.

"Grass" roots vs. "Big Pharma" Patents

I wrote most of this book from my own experiences and the experiences of my patients. We might say that my research came from "The Lab of Life." I also am reading the latest research, including a surprising amount of new books on this subject from MDs, PhDs and other research and clinical scientists. I have also attended some medical meetings on the new Cannabis policies. That is where I heard the term "The Big Experiment."

Big Pharma is trying to isolate individual parts of the plant so they can patent them. (This idea of isolating parts is counterproductive to the patient because the chemicals in combination have a synergistic [68] outcome that is more productive than individually.) Some doctors want books that explain to their patients how to use MM, as this book is doing. But do some physicians and other health professionals want to control that knowledge? I hope this doesn't happen. This movement is already grassroots (pardon the pun) that has started from the ground up (pardon that one too) and amends the single chemical drug model to move into a new paradigm where we become partners in our own healing as described earlier in this book.

I hope (dare I believe?) that this threat of our continued over-attachment to an only-one-drug patented molecule paradigm and the threat of the pharmaceutical industry controlling MM is gradually ending. The now just over half of the U.S. States that are recognizing cannabis for legal and MM use are giving this plant a more natural place in our communities. That expanded awareness leaves us each as an individual now responsible for our own health. This new reality of legalized Marijuana is giving us a chance to grow up instead of Big Gov telling us what to do.

We may not heal as a species until we wake up to our rich mental, emotional and spiritual inner life. Not just by using MM as an aid now, but by using the whole plant that many have known for eons safely and constructively.

Good Drugs, Bad Drugs

There are good drugs and there are bad drugs. Using Marijuana

---

68   Synergy, an abstract concept, is the combination of multiple ingredients producing a result greater than the simple arithmetic summation of the individual components.

is harmful if it is smoked continuously or used in any pattern that brings about a decrease in any important area of our life functioning. Some 9% of users are addicted to it and their life is too often deteriorating which some have said is going 'up in smoke.'
Appendix 8 shows Indicators of Unhealthy or Problematic Use.

It also can be considered harmful in the hands of children and young adults because they haven't matured enough to handle these extreme altered levels of consciousness. How many times in each generation have we seen high school and college students fail classes and drop out of programs partly because they are smoking it and can't focus and learn.

Perhaps more of them are also dependent or addicted to prescribed legal drugs such as stimulants, antidepressants, benzos, alcohol, nicotine, 'mood stabilizers' (expensive and toxic anti-convulsants) and antipsychotics that can complicate making such a diagnosis as cannabis dependence alone.

> Sometimes Pot is the problem.
> Sometimes over-using it is a symptom of an
> underlying but unrecognized problem or trauma.

In the hands of psychologically mature adults who want to expand their understanding of their ill health, for those who are tired of living in denial or being numb, MM can be and often is a helpful natural drug. MM seems to be better used by ideal ratios of THC/CBD from the whole plant and not just as an isolated chemical. And CBD without THC can be helpful on its own. In the mid 1970s MM was helping me significantly when I weighed less than 85 pounds and the plaster cast I was encased in weighed over 30 pounds. MM showed me many things that I would not have known without it. It got me through a difficult period when I needed its help.

I neither support nor overlook all cannabis smoking just because I am discussing MM. If the lungs and breathing tubes are compromised, then it makes sense that it should not be smoked. Vaping is safer and ingesting may be even safer, depending on your individual health needs.

A close friend of mine who had been a heavy tobacco smoker her entire life from age 12 found out she had breast cancer that quickly spread to her lungs and bones. Cancer in a heavy cigarette smoker doesn't necessarily have to start in the lungs. Tobacco use commonly causes almost any cancer. Toxins like tobacco weaken our immune system and the cancer can begin anywhere. So my friend came to me and asked what she should do. Her doctors had previously recommended she stop smoking and recommended that again. I knew she had little time and I just couldn't reason away the idea that she would be suffering with withdrawal from tobacco in her last days. I knew she was dying soon anyway and I didn't want her to suffer any more than she was already.

We did introduce Medicinal Marijuana and her final days were the best any of us could have hoped for. The last time I visited her I brought a CD of Elvis' greatest hits and we danced on and off for two days–while we cooked and laughed.

When her bedtime came I saw how uncomfortable she was with dying. I asked her and she opened up in a way I had never seen her do before in all the 45 years that I had known her because she always kept her feelings and inner life to herself. I listened carefully and we cried together.

I massaged her head, neck and arms. She hadn't been touched in a long time. We talked for a while about the possibility of the continuation of our Soul, our individual existence after death. We had the best time possible for two friends in this situation and I thanked God for helping us. And I still thank God every time I help someone.

I wrote this book to promote the use of MM as a tool to help people heal physically, psychologically, emotionally and spiritually. I neither support nor overlook all cannabis smoking. Like any "drug" MM can help when needed and in some using it may cause problems.

If the patient no longer needs it as Medicinal Marijuana and wants to continue learning from it, I recommend only *occasional* use. If used too often the benefits will fade and what I have observed is the user becoming lethargic and irritable. If using it is treated as a sacrament and used once a week or longer between sessions, it can be what the user wants it to be—a guide to continued learning.

By contrast, if the patient needs MM to help with nausea, anxiety, spasms, and the like, taking in the whole plant (THC plus CBD) may be a daily aid. The "high" that was experienced in the first few days may fade, but the help for the symptoms usually continues.

I know that if I hadn't had a near-death experience all those years ago, being encased in plaster with MM to guide me would have changed me almost as much as the near-death experience did. MM connected me to my natural spirituality. It added to and strengthened my spiritual awareness and growth and thereby helped me find my way back home to the heart of my Real Self connected to God.

Love and Compassion

This is another component of the secret side of MM. It helps us find our own way back to our core, where our compassion lies. This psychological and emotional experience and realization of compassion helps us understand the oneness we can experience with the plant kingdom and all of life. With the right mindset and physical setting it can help us to laugh when we may benefit from it, especially if we may have forgotten how. MM has been available for more than 10,000 years to help us see this bigger Reality.

I wish you the best if you are in a state of dis-ease and trying to heal from it.

I wish you the best if you are being a helper for someone with a problem or illness.

If you are a loved one who is participating in helping someone you love to use this sacred plant, I will close with what my sister-in-law said to my brother while he was dying in ICU. As he apologized to her for all the trouble he had been, trying to recuperate from a heart attack, a stroke and cancer, she said lovingly as she touched him gently:

"Please don't apologize. You and I are in the same experience. We are just playing different roles."

Appendix 1. My Personal MM Use

Until now, we have had to rely on

1) Our prior recreational use experience,

2) The information that we may get from the published literature,

3) Word of mouth from others, or

4) The most experienced plant "pharmacologist" shopkeepers (sometimes called "Budkeepers) in the MM dispensaries where quality and accuracy of information vary from shop to shop.

No matter what we may find, we should still use a personal trial and error method as we ingest MM each time, while keeping an accurate diary or log of MM type, delivery method, dose and result for each problem for which we are using it. Doing so will help you and us (if you ever share these details) to have a better understanding of what works for you and what does not.

Using Trial and Error

All of the above helps to explain how and why the US States and countries that have enacted MM provisions or decriminalized using cannabis have created a double edged sword result:

• Using cannabis (or its selected components) carefully may help lessen some peoples' problems.

• But there are no agreed-upon guidelines from accepted peer-reviewed medical groups.

Most States and countries used a common *political process* to enact these MM laws: a combination of grass roots patients and those in their support system convincing politicians to propose a popular vote on MM and making the results of an acceptable majority into a state law. Some States, such as Georgia and Louisiana (both now beginning a kind of 'MM light'), enact MM laws within their congressional process and do not use a public vote.

If your state enacts an MM law, as an individual you may be left with only one possibility: find a specially licensed or certified physician and convince them that you have an acceptable MM-responsive ailment to write you a positive form to submit to a state authority. See the USA States map on page 40 for our current growth.

Appendix 2. *(cont'd from page 122)*

Table A 1. My Personal MM Use –
A Trial and Error Results Log

Date	Problem	MM Type	Method	Dose	Results
10 Oct 15	Fear and anxiety*	CBD drops	Deposit under tongue or vape oil	5 to 15 drops 1-2 x/day**	Quick calm

*Assuming otherwise treatable causes are ruled out.

** See instructions in drops literature.

Appendix 3. Tart Classic *(cont'd from page 26)*

In his classical and then groundbreaking 1971 detailed study of 150 cannabis users' experiences when 'stoned,' Charles Tart PhD found enough information to fill an outstanding book.
Of those, nearly all were college students and 86% were 17 to 30 years old. It was and still is a remarkable report.
Below is one of many useful tables from that book.

Table A 2. 'Good' vs 'Bad' Cannabis Effect
VALUES OF VARIABLES FOR MAXIMIZING
PROBABILITY OF 'GOOD' OR 'BAD
EFFECT' from Tart et al 1971 [69]

	VARIABLES	GOOD Effect LIKELY	BAD Effect LIKELY
Drug	Quality	Pure, known.	Unknown drug or unknown degree of (harmful) adulterants.
	Quantity	Known accurately, adjusted to individual's desire.	Unknown, beyond individual's control.
Long-term factors	Culture	Acceptance, belief in benefits.	Rejection, belief in detrimental effects.
	Personal-ity	Stable, open, secure.	Unstable, rigid, neurotic, or psychotic.
	Physiology	Healthy.	Specific adverse vulnerability to drug.
	Learned drug skills	Wide experience gained under supportive conditions.	Little or no experience or preparation, unpleasant past experience.

69   Tart C 1971 *On Being Stoned*: A psychological study of marijuana intoxication. Science and Behavior Books

VALUES OF VARIABLES ...cont'd

	VARIABLES	GOOD Effect LIKELY	BAD Effect LIKELY
Immediate user factors	**Mood**	Happy, calm, relaxed, or euphoric.	Depressed, overexcited, repressing significant emotions.
	Expecta-tions	Pleasure, insight, known eventualities.	Danger, harm, manipulation, unknown eventualities.
	Desires	General pleasure, specific user-accepted goals.	Aimlessness, (repressed) desires to harm or degrade self for secondary gains.
Experiment or situation	**Physical setting**	Pleasant and esthetically interesting by user's standards.	Cold, impersonal, "medical," "psychiatric," "hospital," "scientific."
	Social events	Friendly, non-manipulative interactions overall.	Depersonalization or manipulation of the user, hostility overall.
	Formal instruc-tions	Clear, understandable, creating trust and purpose.	Ambiguous, dishonest, creating mistrust.
	Implicit demands	Congruent with explicit communications, supportive.	Contradict explicit communications and/or reinforce other negative variables.

Appendix 4. Marijuana, codeine *may* be equal as analgesics

Pain Medicine *Clinical Psychiatry News* 2004

Positive results: 19 of 25 (76%) of studies and 2/3 of randomized trials

VANCOUVER BC -- Marijuana is an effective analgesic that probably can be equated with codeine in its potency, Dr. David Boyd said at the annual meeting of the American Pain Society.

Cannabinoids are difficult to study in a placebo-controlled clinical trial because patients can tell when they get active drug, but Dr. Boyd said that his review of the literature identified 25 human studies of its use as an analgesic, of which 19 reported positive findings. Of those studies, 12 were randomized clinical trials, of which 8 reported positive findings.

The evidence suggests that marijuana and its cannabinoid derivatives "are going to be a useful adjunct in some patients," said Dr. Boyd of the University of Western Ontario, London.

The indications in these studies for which marijuana or a cannabinoid derivative was used included postoperative, cancer, and neuropathic pain. Most of the trials were small, with an average of 10-20 patients; the largest trial had 40 patients.

In terms of analgesic effect, cannabinoids and codeine may be roughly equivalent, based on the positive studies.

The evidence suggests that cannabinoids have a synergistic effect when used with opioids and capsaicins. There also may be a synergistic effect when they are used with anti-inflammatories and with muscle relaxants, Dr. Boyd suggested.

Medical marijuana is an issue that elicits both zealousness and controversy, he said. But there is nothing new about it. Analgesics have long been made from plant sources, cannabis included, and the identification in 1988 of the endogenous cannabinoid system, which functions in parallel to the opioid system, supports this use.

Cannabis has been used medicinally for at least 1,500 years, and in the 1800s it was quite popular as an analgesic. Queen Victoria reportedly used cannabis for menstrual cramps, noted Dr. Boyd, who said he has become open to using it in his practice. Two cannabinoid products, nabilone and dronabinol, are legally available in Canada.

Moreover, patients with pain appear to use it quite commonly, at least in Canada. A recent survey of chronic pain patients in Montreal reported that 15% acknowledged using medicinal cannabis for pain relief. [70]

---

70   Kirn T: Marijuana, codeine may be equal as analgesics. (Pain Medicine) Clinical Psychiatry News (Magazine/Journal) July 1, 2004 Intl Medical News Group Volume: 32 Issue: 7 Page: 65(1) (Dr. Boyd received funding to conduct his review from ICN Canada, which distributes nabilone.)

Appendix 5. Hands-on Healing Meditation

This type of hands-on energy work is easy to use with people who have a chronic condition or those making their transition. The beauty of this work is that the loved ones who have been afraid to touch the patient now have a way to give their love physically. And the one receiving is given the gift of a deeper peace.

Let the receiver pick their favorite soft music with no words and play softly in the background. The one receiving the energy lies down and is usually covered with a light blanket. Other members of the family, friends or health care team (whom the receiver feels are "safe") gather around and sit with their hands gently and lightly touching the receiver.

Before we begin, I always say a prayer to connect and unite us all in what we are asking to do. I say something like," Dear God (or Dear Holy Mother, or with Whom my patient feels most comfortable) please may we be instruments of your healing energy and your Oneness. Please help us to get our egos out of the way so You may come through."

This works with one person as an instrument, or as many as six. The main principle is that the people that are doing the hands-on-healing feel safe to the patient receiving the energy. I have worked with children as young as three in a healing circle, and adults as old as 90 who may be dying themselves. In one such case, when we had finished the first "healing", we covered that patient with another blanket, and then a 90-year-old who had been giving energy with us, laid down on the floor a few feet away, and we repeated the healing on her.

A 5-year-old granddaughter asked us to do it to her. Her mother called me a few weeks later to tell me that she was going through a painful divorce and since it had started, her little girl had stomach pain to the point where they had taken her to a medical specialist for tests. The tests were normal. And, since we did the healings that day her daughter has been pain free. This mother told me she was in school to become an acupuncturist and after experiencing that day with the healings she now believed more than ever in energy work.

It doesn't matter if the people giving the healing believe in all this. All that matters is that they have the intention of wanting to help. Many times the person receiving healing tells me afterwards that their pain medication is now working or they no longer feel they need it. Their coloring improves, they feel more relaxed and they feel loved.

Hands-on-healing can be done twice a day, or more, if the patient requests it. The givers have often told me that they calmed down from the experience. It not only helped their bodies to relax, it also helped their hearts to know that they were giving "something" to the person in need. This is especially important for loved ones who may have been afraid to touch the patient for fear of hurting them. This gives them a safe way to express themselves that bypasses words.

Whether the patient is going to get better or is going to die, this doesn't matter in these hands-on circles. We are not looking for a "cure." We are looking for a sharing that brings comfort to the receiver, and usually spills over into the givers.

We do this together for about 20 minutes. It feels like a meditation. We clear our minds and sit in a peaceful way, placing our hands gently on the patient and sharing a current of Energy that envelops all of us. We naturally come out of it within 20 minutes or so, feeling more relaxed and at the same time energized, because the energy we shared didn't come from us but through us.

We close this healing circle with a prayer of gratitude: "Dear God, Dear Spirit, thank you for allowing us to be instruments of your healing Energy. Thank you for allowing us to feel your Oneness. ... Amen"

Appendix 6. More Information on MM and Cannabinoids

Cannabinoid basics part 2

Cannabinoids are a group of various chemical compounds found in the cannabis plant which act on the receptors located on cells which repress the release of certain neurotransmitters (NTs) in the brain. Different strains may have varying levels of cannabinoids and affect dosing when used for medicinal purposes.

When found in nature, cannabinoids are present in the form of sticky resinous structures known as glandular trichomes. A single trichome gland will contain a variety of cannabinoids, and recent research has shown that these cannabinoids work together to produce their medical benefits. This phenomenon has been called the "entourage effect" because it shows that a diverse variety of cannabinoids, in precise ratios, has a more significant medical benefit than an isolated single cannabinoid synthesized in a lab.

The Plant Species

Although this information is a bit technical, it is important. Please bear with me here. Some of it has practical usefulness.

Cannabis sativa, C. indica, and C. ruderalis are all species of the genus Cannabis. They can all inter-breed freely, and many 'pedigree' cultivars are indica/sativa hybrids.

Although many authorities continue to class all plant varieties, including hemp and marijuana, as Cannabis sativa, most accept that there are three separate species or sub-species.

C. sativa, the most widely cultivated in the Western World, was originally grown on an industrial scale for fiber, oil, and animal feedstuffs. It is characterized by tall growth with few, widely spaced, branches. C. indica, originating in south Asia, is also known historically as Indian hemp, and is characterized by shorter bushy plants giving a greater yield per unit height. C. ruderalis is a hardier variety grown in the northern Himalayas and southern states of the former Soviet Union, characterized by a more sparse, "weedy" growth and is rarely cultivated for its THC content.

In nature, there are two major cannabinoids and they are ancient in their existence and function. The first are found within our own bodies, including our brain and nervous system. It is called the endo-cannabinoid system, abbreviated as eCB for endo-cannabinoid and eCBS for the endo-cannabinoid system.

Cannabinoid Receptors

We are all born with CB1 cannabinoid receptors on the cell membranes in our brain and nervous system and CB2 receptors in these and in certain cells of our immune system and other organs. These are basic parts in our body's complex and powerful *endocannabinoid system* introduced in chapter 11 above. The now-called **eCBS** was discovered in 1988 through 1992-4 by Howlett, Devane and Mechoulam and their colleagues (described in the 2014 *Handbook of Cannabis* and many authoritative places elsewhere). This eCBS was recently found to be the largest receptor system in our body.

CB**2** receptors appear to be responsible for the ability of THC, CBD and the terpenoids, aka flavonoids, to reduce inflammation, some pain and other health problems. CBD also has a unique psychoactivity by its anti-fear, anti-anxiety and common calming effect when taken sublingually or inhaled in an oil as vapor when they bind with the CB2 receptor. Both CB1 and 2 receptors are also called "G protein-coupled receptors" and are found in the plasma membrane of cells. [71] Table A 3 below summarizes some of their differentiating and similar characteristics.

Table A 3. Cannabinoid (CB) Receptor Characteristics

CB Receptor	1	2
Binds	THC and Anandamide	CBD (cannabidiol)
Receptors located in	Essentially all organs	Eyes, heart, GI tract, pancreas, bone, immune system, brain
Plant binding	Sativa binds to both CB1 and CB2	Indica tends to bind to CB2
***Both* receptors are *activated* by** *Anandamide (AEA) and 2-AG function as diffusible and short-lived intercellular messengers that modulate nerve synaptic transmission	• The Endocannabinoid NeuroTransmitters (NTs): 1) Anandamide and 2) 2-AG (2-arachidonoyl glyceride), • plant cannabinoids, such as THC, CBD and •synthetic THC analogues	

71   Summarized from Pertwee 2014 *Handbook of Cannabis*; Mechoulam & Parker 2013; and en.wikipedia.org/wiki/Cannabinoidreceptor

Unknown until discoveries in 1988 through 1992/4, the CB1 receptor has been identified as the *most prevalent neurotransmitter system* in the human brain, appearing also in the cerebellum and reproductive systems of both genders. CB2 was originally thought to appear mostly in the immune system but later it was found to function in many organs, including our brain. In addition to other actions, it is likely responsible for cannabis' anti-inflammatory effects.

Classic Psychoactive Cannabinoids

The four psychoactive cannabinoids include Δ9 (delta-9) **THC** (tetrahydrocannabinol), Δ9 THCV (tetrahydrocannabi-varin), Δ8 THC, and CBN (cannabinol). Especially THC, their actions produce progressively dose-related effects on at least seven of our mental, emotional and physical life experiences and capabilities. These experiences include our perception (surface sense, 'take' or observation), cognition (thinking, reasoning), emotion (feelings, 'temperature'), mood (background feeling, 'weather'), memory (decreased working/short term), reflexes (altered motor function), sensory experiences (increased hearing, seeing, appetite, taste, touch, sexual response).

With increasing doses these effects can be varied enough to span a range from 1) enjoyment and fun to 2) progressive dysfunction to 3) being embarrassed if in unsafe company to 4) having a psychedelic experience (pleasant or unpleasant) to 5) paranoia to 6) danger if driving a vehicle or operating machinery. This is why an inexperienced user should be cautious and go slow with the amount of their intake and dose. An important aspect of dosing has to do with the ratio with these above Psychoactive THC chemicals to the classically more non psychoactive but still medically effective cannabinoids found in the specific natural plant strain used. Depending on the THC/CBD ratio used, after several days in a row of taking MM the degree of a 'high' will no longer be there from the THC but the other medicinal benefits will usually continue. Expanding on the 'THCs,' I describe these in more detail below, starting with Table A 4.

Table A 4. Four Psychoactive Cannabinoids Characteristics *

Names	Comments	
Δ⁹ THC Tetrahydrocannabinol **18 types** found to date*	Most psychoactive cannabinoid. See Apendix12 &13 p 181 & 187	Needed combined with CBDs for several problems and illnesses
Δ⁹ THCV Tetrahydrocannabivarin # types not found a propyl form, needs heat to form THC	A homologue (slight variant) of THC and more found in sativa	Anti-seizure, anti-diabetic, stimulates appetite & bone growth. See Table A 10 on p 183
Δ⁸ THC 2 types May need development as a separate drug or for combined use	Very low doses (0.001 mg/kg) caused increase in appetite and possibly clearer thinking*	Possible therapeutic agent in the treatment of weight disorders [72]
THC-A prominent THC precursor acid; when heated → psychoactive THC; also a Δ9		
CBN cannabinol Number of types not found *from chaps 1 & 6 *Handbook of Cannabis* 2014	Weakly psychoactive; mostly a *THC metabolite*; found in trace amounts; a weak agonist of CB1 receptors Higher affinity to CB$_2$ receptors	

CBDs: Classic Non-Psychoactive Cannabinoids

Cannabidiol (CBD) is one of at least 85 active cannabinoids that have *not* been *detected* in *any other plant*. It is a major phytocannabinoid (*phyto* means *plant)*, accounting for up to 40% of the plant's extract. While not negating THC's therapeutic power, CBD appears to have more medical applications than THC. Though cannabidiol is usually said not to be psychoactive, that idea is in large part a misnomer. This is because CBD's effect is • *calming* and thus • mildly *psychoactive* in that important respect. And it acts to • counter the strong 'high' effect of THC. In contrast to benzodiazepines, antidepressants and similar psychiatric drugs, CBD's calming effect is neither numbing, unsettling nor clouding of our awareness. It is often helpful for PTSD, insomnia and opiate, benzo and other drug withdrawal.

CBD concentrations range from none to rarely 95 percent of the total cannabinoids present. THC/THCA and CBD are the two most abundant naturally occurring cannabinoids. For CBD to affect the high, THC must be present in quantities ordinarily psychoactive. CBD can contribute to the high by interacting with THC to potentiate (enhance) or antagonize (interfere or lessen) certain qualities of the high. CBD appears to potentiate THC's downer effects and counter its upper effects. While CBD delays the onset of the high it can make it last longer (as much as twice as long) which can be slow to come on but often 'keeps coming on.'

An orally-administered liquid containing CBD has received orphan drug status in the US for use as a treatment for dravet syndrome and other rare kinds of epilepsy, under the brand name Epidiolex, with 98% CBD and "no THC." [72]

---

72   Summarized from Pertwee 2014 *Handbook of Cannabis*; Mechoulam & Parker 2013; and en.wikipedia.org/wiki/Cannabinoidreceptor

But CBD oil and tincture are already legal in all 50 states and available as generics by various makers on Amazon.com and at some local health supplement stores at a reasonable price. I experienced complete relief for up to 12 hours using CBD (500mg in 2 oz glycerol oil) of PTSD related chest pain and sense of dread or angst. It also relieved some of the arthritis pain in both of my knees. It works better than the prescribed toxic muscle relaxant drugs and helped me sleep. I found that vaping the same CBD oil works quickly and it may work for many 500 mg CBD oils used in an ordinary new 'nicotine' type vaping CE4, etc, device.

Sativex - Cannabidiol is made in the UK, etc in a 1:1 THC:CBD ratio as a sublingual spray marketed as Sativex (generic name nabiximols). It comes as an oro-mucosal spray derived from two strains of cannabis leaf and flower, cultivated for their controlled ratio of the active compounds Δ9 THC and CBD.

Sativex is used for symptom improvement in adults with moderate to severe spasticity due to multiple sclerosis (MS) who have not responded adequately to other anti-spasticity medication and who have clinically significant improvement in spasticity related symptoms during an initial trial of therapy. It has also been used to lessen the pain associated with advanced cancer. Treatment with Sativex must be initiated and supervised by a physician with specialist expertise in treating this patient population.

Sativex is so expensive that many have and will likely continue to use plain cannabis instead for their pain. Patients report that *Care By Design* CBD 1:1 is helpful for neuropathic pain, drug withdrawal, traumatic brain injury, MS and arthritis. The amount of CBD and THC is reported as 16 mg of CBD/ml in a 2.6 mg/spray. Buying it appears to be limited to CA residents with a MM state issued certificate. I don't know if other MM approved states sell it. I have not used Sativex. Keep researching online for new products and results.

For most MM or recreational users the only way to know the results of using any cannabis is through trial and error (and this principle is *true for all* FDA and DEA approved drugs). Few MM sellers and dispensaries will have accurate analysis results available for buyers to use to decide what to buy. Even so, I have included enough information about these cannabinoids here in the text and tables to have a basic understanding about MM and how to begin using it (summarized in this chapter expanded from what one lab and dispensary publish online).

CBC (Cannabi**chromene**) is probably not psychoactive in pure form but some suggest that it may interact with THC to enhance the high, and is thus *indirectly* psychoactive. Others say that it does not impact THC's psychoactivity. It may play a role in the anti-viral and anti-inflammatory effects of cannabis, anti-proliferative (cancer fighting) effects and could improve the analgesic effects.

CBG Cannabi**gerol** is also non-psychoactive and usually in high concentrations in hemp. It may decrease intraocular pressure in glaucoma and have antibacterial effects. It may alter the overall effects of cannabis on users. I summarize these and more in Table A 5 on the next page.

Table A 5. Five Classically Non-Psychoactive Cannabinoids and the Terpenoids *

Names	Comments	
CBD cannabidiol	The most available non-THC cannabinoid in concentrated hemp extract by tincture or decarboxylated and filtered CBD oil. It is essentially non toxic.	Increases some of the effects of THC while decreasing others (see text). Calming and thus mildly and *effectively* psychoactive.
CBC cannabichromene	Probably not psychoactive See text	
CBG cannabigerol		
CBN Cannabinol	Breakdown products that comes with age, light and heat. Sedating.	
CBL cannabicyclol		
Terpenoids limonene, myrcene, a-pinene, linalool, β-caryophyllene, caryophyllene oxide, nerolidol and phytol	Share a precursor with cannabinoids. Are common flavor and fragrance components in human diets. Recognized as Safe by the US FDA	Cause the unique odor of burned cannabis, as THC and CBD don't smell. Display unique therapeutic effects that may contribute meaningfully to the 'entourage' effects of cannabis-based medicinal extracts

* summarized from R Pertwee *Handbook of Cannabis* 2014 chaps 1 and 7

CBN (Cannabi**nol**) is produced as THC ages and breaks down through a chemical process called oxidization. As opposed to CBD, high levels of CBN tend to make the user feel confused rather than high. CBN levels can be kept to a minimum by storing cannabis products in a dark, cool, airtight environment. The herb should be dry prior to storage and may have to be dried again after being stored in a humid location.

CBL (Cannabi**cyclol**), like CBN, is a degradative product. Light converts CBC to CBL. If you are a grower, you can experiment with different strains to produce the various qualities you seek. A medical user looking for something with sleep-inducing properties might want to produce a crop that has high levels of CBD. Another user looking for a more energetic high will want to grow a strain that has high levels of THC and low levels of CBD. In general, cannabis sativa has lower levels of CBD and higher levels of THC. Cannabis indica has higher amounts of CBD and lower amounts of THC than sativa.

Terpenoids aka Flavonoids typically are volatile molecules that evaporate easily and readily announce themselves to the nose. They produce a synergy with cannabinoids when used to treat pain, inflammation, depression, anxiety, addiction, epilepsy, cancer, fungal and bacterial infections. Ethan Russo MD and others showed scientific evidence for these non-cannabinoid plant components as assumed antidotes to intoxicating effects of THC that could increase its therapeutic index (i.e., make it work better). He suggested methods for investigating 'entourage' effects in future experiments on MM. If proven, this kind of cannabinoid-terpenoid chemical collaboration would increase the likelihood that a pipeline of new cannabis therapeutic products is possible. [73] Here similar terms as synergy, entourage effects, chemical collaboration and additive effects further define how these three cannabis components of THC, CBD and terpenoid/flavonoids work together.

Patients who abandon a suitable strain for one with higher THC and/or CBD content may not get more relief if the terpenoid profile is significantly different. Some 200 terpenes have been found in cannabis, but only a few of these odiferous oily substances appear in amounts substantial enough to be noteworthy or nose worthy. The terpenoid profile can vary considerably from strain to strain (note: there are an estimated 2000 different strains of cannabis.) Terpenes and CBD buffer THC's remarkable psychoactivity.

Like their odorless cannabinoid cousins, terpenes are oily compounds secreted in marijuana's glandular trichomes. Terpenes and THC share a biochemical precursor, geranyl pyrophosphate, which develops into the cannabinoids and terpenoids that saturate the plant's flower tops and spackle its leaves. [74] Cannabinoid-terpenoid interactions amplify the beneficial effects of cannabis while reducing THC-induced anxiety and they do more as reflected in the rich cannabinoid scientific and clinical literature, some of which I reference in the footnotes and reference section.

---

73   Russo EB 2011 Taming THC: potential cannabis synergy and phytocannabinoid-terpenoid entourage effects *Brit J Pharmacol* 163(7): 1344
74   From Gardner and Lee accessed 10 Nov 2014 at projectcbd.org/ medicine/ terpenses/

Summary and Conclusions

Big Picture

THC has double edged sword qualities. It has to be used carefully. Too much of it or a too-high ratio to CBD can cause acute problems with: *thinking*, *remembering*, *communicating*, *anxiety*, *balance* and *overall functioning* (a 'THC overdose' or toxicity). But taken in the right dose for the right symptom at the right time it can help selected symptoms and problems.

By contrast, CBD is usually a single edged sword, helping to lessen more symptoms with essentially no toxicity.

A **Little Picture** but always important view of using MM is to find the right dose for both THC and CBD, as described in Appendices 11 and 12 on page 178 through 184.

*   *   *

MM Decisions Review

This ends this important section. If you have time or interest in making decisions regarding your MM use, I suggest that you read and study Appendix 11 on Useful *Scientific* Information in Decision Making when Using MM on page 178 and then Appendix 12 on Useful Scientific and *Technical* Information in Decision Making when Using MM on page 181 below.

Then before you begin, go back up to pages 120-122 in chapter 11 to **Cannabis Dosing Guidelines** to finalize your first MM plan.

Appendix 7. New Light on the Darkness of PTSD... cont'd from p 100

"Concentrated in the brain and central nervous system, the cannabinoid receptor known as CB-1 mediates a broad range of physiological functions, including emotional learning, stress adaption, and fear extinction. Scientists have determined that normal CB-1 receptor signaling deactivates traumatic memories and endows us with the gift of forgetting.

But skewed CB-1 signaling, due to endocannabinoid deficits (low serum levels of anandamide), results in impaired fear extinction, aversive memory consolidation, and chronic anxiety, the hallmarks of PTSD.

PTSD is one of many enigmatic conditions that may arise because of a dysfunctional endocannabinoid system. A 2009 report by Virginia Commonwealth University scientists discerned a link between the dysregulation of the endocannabinoid system and the development of epilepsy. Researchers at the University of Rome in Italy have documented low levels of anandamide in the cerebrospinal fluid in patients with untreated newly diagnosed temporal lobe epilepsy.

Dr. Ethan Russo postulates that clinical endocannabinoid deficiency underlies migraines, fibromyalgia, irritable bowel disease, and a cluster of related degenerative conditions—which may respond favorably to cannabinoid therapies.

Individuals have different congenital endocannabinoid levels and sensitivities that factor into how one responds to stress and trauma. Alcoholism induces endocannabinoid deficits. So does lack of exercise and a diet laden with corn syrup and artificial sweeteners." [75] ... this ends PTSD article

*   *   *

420 - What does it mean?

Let's look at both a little history and have some fun here.
Summary: 420 is a code word.

Rob Griffin editor of 420 Magazine wrote "One thing is certain to me, Brad Bann (aka 'the Bebe' in San Rafael high school [in 1970]) coined the term 420 and 'the Waldos' carried the term across the U.S. on tour with the Grateful Dead. I took the torch in 1993 and promoted 420 to the world via my website/s, reaching over 20 million people a year, totaling over 420 million people worldwide. Now there are billions of us.

"People have asked me the same question, 420 million times, 'What is 420?' The most common reply was usually an hour long explanation of 420 different things that it is and can be. After 20 years of promoting this magical number, I've come to summarize it down to, 'It's anything you want it to be.' ... Sociologists have weighed in with their 'expert' 420 viewpoint (and they get paid for this!?) of what it means. '420 creates an intense sense of group belonging among friends, strangers, and crowds' or 'a kind of mystical, spiritual, or extraordinary sense of belonging, where the group exists as a reality greater than itself' ... What? Way too stoned in Madagascar I'm afraid."

---

75   MAPS Annual Report maps.org/news-letters/v23n3/ v23n3_40-42.pdf

Appendix 8. Unhealthy or Problematic Use Indicators

Answer 'Yes' or 'No' to each of the following questions.

John Craven MD suggests that if you answer yes to more than a few consider consulting a professional caregiver.

Do you feel compelled to use marijuana when you do not need to do so?

Do you often use more marijuana than you had intended?

Do you lose control over your use - after you take some marijuana?

Have friends or family expressed concern about your use of marijuana?

Do you feel guilty or 'wrong' about your use?

Do you often use marijuana for no personal medicinal purpose?

Has your use of marijuana resulted in problems in your home?

Has your use resulted in medical or health problems?

Do you experience rebound withdrawal symptoms if you do not use every day?

Have you had problems in your workplace as a result of your use?

Have you ever or do you now abuse prescription or other drugs?

Do you take other psychoactive drugs at the same time as you use marijuana?

Do you drink alcohol at the same time as you use MJ?

Have you run into financial problems as a result of your use?

Do you become angry or instigate fights while under the effect of marijuana?

If you travel internationally - do you take unnecessary risks in using marijuana?

Craven concludes: "The nice thing about marijuana - compared to benzodiazepine or opiate drugs used for similar conditions of anxiety, sleep sedation or pain - is that if you run into trouble with your use - you can stop using without serious medical complication, difficult or dangerous withdrawal. If you want to reduce or to stop using but find that you cannot do so - you are best to seek professional help or attend to a community based recovery support group - where you will find people who have been where you are - and who are willing to share their experience in recovery with you." [76]

Some use the *Michigan Alcoholism Screening Test* and answer the questions for MJ instead of for alcohol.

---

[76]   Craven J (2014-04-12). *The Power In Pot*: How to Harness the Medicinal Properties of Marijuana in the Management of Clinical and Stress Related Conditions. (Kindle Locations 2428-2441). supportnetstudios.com

Fig A 5 Spectrum of Psychoactive Substance Use

From Every Door Is The Right Door: a British Columbia planning frame-work to address problematic substance use and addiction. May 2004. Aggarwal et al. *Harm Reduction Journal* 2012 9:4 doi:10.1186/1477-7517-9-4

"In fact, the whole substance use/abuse dichotomy ought to be discarded and the transition be made to a spectrum view, as has been adopted by the British Columbia Ministry of Health. In their framework for addressing problematic substance use they include the diagram below (see Figure) and note: 'The Framework recognizes that instances or patterns of substance use occur along a spectrum from beneficial use to non-problematic use to problematic use (including potentially harmful use and substance use disorders). Substance use disorders represent the extreme and most damaging end of the spectrum. Some people choose to abstain from using psychoactive substances while some people choose to use only certain substances. It is important to emphasize that abstinence is a healthy lifestyle option. Nevertheless, many people choose to use substances and some do not develop serious problems because of this use.'"
(page 8)

Appendix 9.

Table A 6. Voice of the ego vs. the Real Self/Soul

Ego Traits	*Real Self/Soul Traits*
Becomes bored easily	Realizes peace when doing nothing
Commands	Suggests
Demands	Guides
Tests	Nudges
Chooses for you	Leaves choice to you
Imprisons	Empowers
Promotes dependence	Promotes independence
Suffering – "why me?"	"Pain is part of life"
Intrudes	Respects
Pushes	Supports
Excludes	Includes
Instills fear	Promotes well-being
They	We
Is status oriented	Is free and open
Judges	Accepts Individuality
Demands obedience	Encourages growth and development
Implies having ultimate authority	Recognizes a Higher Power
Offers shortcuts	Offers integration
Seeks personal gratification	Extends Unconditional Love
Self-righteous	Humility
Creating, maintaining and defending boundaries	Dissolving boundaries

Barbara Harris RT

Appendix 10.

**Timeline of Cannabinoid Discovery and MM
Movement Development**
from 1889 to Present

From the Beginning of Earth Time: Please consider these two basic questions -

1) How would we humans and other animals have developed these two major natural cannabinoids (called **endo**cannabinoids) and receptors for them and for exogenous cannabis plant chemicals in our Central Nervous System (CNS), brain, nerves and most body organs? **2**) Why would God or Nature make the cannabis plant to survive over the countless millennia?

Here is a timeline documenting what and when we have found about answering that question.
I also show in bold *italics* how some groups have not understood the meanings of these natural things.

1889 CBN (cannabinol) first compound **isolated** in pure form [Wood, 1899] initially wrongly assumed to be the main active compound of the plant responsible for its psychoactive effects [Mechoulam and Hanus, 2000]

1906 FDA Food and Drug Administration starts to regulate the labeling of products containing alcohol, opiates, cocaine and cannabis and other substances (a good idea that has gone bad over last 40 years).

1937 Marijuana **Tax** Act began a slow government control process over a mistaken belief it had no medical benefit.

1963 CBD (cannabidiol) second cannabinol found [Mechoulam and Shivo] [77]

1964 THC isolated as the main active cannabis compound, by Gaoni & Mechoulam

1970 *Controlled Substances Act* Psychoactive drugs 'scheduled,' with MJ as schedule 1 ('no medical use' and authorities believed had a 'high abuse risk')

1972 Richard Nixon appoints Shafer Commission that urged cannabis be re-legalized. Their recommendation was ignored by the FDA and other authorities

1973/4 *DEA* Drug Enforcement Administration begins

1974 *NIDA* National Institute on Drug Abuse still today holds monopoly by controlling all cannabis research

1975 FDA establishes 'Compassionate Use program' for medical marijuana

1980 420 molecules found in cannabis by Turner et al Since then, over 530 found and more [77]

1988 THC binding sites identified in the brain [Devane et al], which began finding the ECS [77]

1990 CB 1 R cloning [Matsuda et al. 1990] named the 'Cannabinoid Receptor System' (CRS) due to the binding affinity of THC to these receptors as partial agonist

1992 Endocannabinoid system confirmed (based on above CRS) by WA Devane et al and 'Anandamide' (N-arachidonoylethanolamine or AEA), the second found endocannabinoid neurotransmitter (NT). [Mechoulam et al]

1993 CB 2 R second cannabinoid receptor discovered by Munro et al [77]

[77]   McPartland JM, Russo EB 2014 Chapter 15. Non-phytocannabinoid constituants of cannabis and herbal synergy. In Pertwee *Handbook of Cannabis*

1994/5 2AG (2-arachidonylglycerol) described as another endocannabinoid neurotransmitter

1996 Compassionate Use Act of California starts MM State movements

1997 Medical Utility of Marijuana Workshop, NIH

1999 National Academy of Sciences, Institute of Medicine Report medicalmarijuana.procon.org/ view.answers.php?questionID=255

2000 CMCR (Center for Medicinal Cannabis Research) first medical school cannabis research U Cal San Diego

2003 HHS of USA takes out a **patent** on medical cannabis with a mix of theory and unclear purpose

2008 THC, as a partial agonist, resembles anandamide in its CB1 affinity, although with less efficacy [Pertwee]

2009 Medicalization of Cannabis wellness seminar

2015 24 States and DC have MM and/or legalized cannabis use; the Republican congress makes illegal for fed govt to overrule or harass states MM laws and delivery and Feds are forced to agree (see page 40)

Thousands MD/PhD level investigators & professionals continue to discover and publish on new basic and clinical cannabis and MM information (see e.g., *Handbook of Cannabis*)

17th c to present compressed cannabis trichomes called hashish remains a world trade [78]

2015 National Clinical Conference on Cannabis Therapeutics in USA continues see medicalcannabis.com

---

78   See also erowid.org/plants/cannabis/cannabis_timeline.php

Barbara Harris RT

Appendix 11. Ten Variables to Address When Using MM

Throughout this book I have discussed how important it is to pay careful attention to several factors when using MM. As should be clear by now, this plant is not a single chemical or drug as we have been used to taking when our physician prescribes one or when we get it over the counter.

In this section is a summary and review of the various attentions that we need to assure to make MM work for us. In Table A7 on the next page I list all ten of these basic variables and comment on each. After the table I have reproduced an identical table on page 180 with large spaces to record your personal responses or comments for when you use MM. Make several extra copies for you to use in the future if you may need them.

Table A 7. Ten **Variables** to **Address** When Using MM

Variable	Comments
Cannabis Strain	Covers a spectrum from sativa/indica/hybrids to genetic makeup ('blueprint for growth'/genotype) to physical characteristics (phenotype) which underlie countless strain varieties. See Kindgreenbuds.com and other sources.
THC/CBD Ratio	Ratios range from high THC/low CBD (easiest to find) to the reverse (hardest to find).
THC/CBD % of each	Both % and mg amount are important and useful to know if you can find that information.
THC/CBD mg each	Record these numbers carefully in your journal in your trial and error process.
Delivery method	Can use trial and error to find which method works best.
Set (mental)	Identify and record set and setting in your journal for current and future use.
Setting	Be sure setting is safe and supportive.
Problem used for	Record name of problem on a scale of 1 to 10 before, during and after using MM. Also record same 1 to 10 scale separately when not at all 'stony' for rating your response to the MM.
Safety	Assure that you and others around remain safe during your MM use.
Individual Sensitivity	Know that people respond differently to the same strain and dose of MM. Record your results every use.

*See Table A 8 below to use to record notes on your experience and response to using MM.

Table 8. My Notes on Ten Variables to Address
When Using MM

Variable	Comment
Cannabis Strain	
THC/CBD Ratio	
THC/CBD % of each	
THC/CBD mg each	
Delivery method	
Set (mental)	
Setting	
Problem used for	
Safety	
Individual Sensitivity	
My Personal Result for this session	

Appendix 12. Cannabinoid Basics 3

Useful Scientific Information in Decision Making when Using MM

What else do we know about the potential of using MM? Here are three tables that I have compiled from the basic and clinical science literature on cannabis, which I introduce in the next three paragraphs.

For understanding Table A 9 - In their comprehensive evaluation of the clinically and practically important THC and CBD effects when using MM, physicians Ethan Russo and Geoffrey Guy compiled the following table from 17 peer-reviewed published articles in scientific and clinical journals. As an MM user or as an assistant you can review these effects of THC and CBD to help make decisions for yourself in Table A 9 on the next page.

Study that table a minute. And then look at the next paragraph which introduces a similar and more detailed Table A 10 as shown on the page following Table A 9. For more on THC effects see Appendix 13 on page 187-9.

For Table A 10 - Expanding the above MM related information in A 9 that you just studied and from the Pie Chart on page 125 in Chapter 11 above, the director of CannLab in Denver, CO compiled the next table based on 74 peer-reviewed published scientific and clinical studies available in 2012. These reports were written by MDs, PhDs and other researchers and published in respected scientific and clinical journals from 1975 through 2012. This table may give the patient and caregiver more detailed information not previously readily available.

Table A 9. Effects of THC and CBD updated *
++ = most; + = effect; ± = some; − = no

Effect	THC	CBD
CNS Anticonvulsant	+	++
Muscle relaxant	++	+
Anxiolytic	±	++
Psychotropic	++	−
Antipsychotic	−	++
Short-term memory problems	+	−
Distortion of perception of time	++	−
Neuroprotective antioxidant	+	++
Antiemetic	++	++
Sedation	+	−
Cardiovascular		
Bradycardia	−	+
Tachycardia	+	−
Hypertension	+	−
Hypotension	−	+
Appetite/GI/metabolic		
Appetite	+	−
GI motility (slowed)	++	+
Metabolic/diabetes	+	−
Anti-cancer		
Glioma (apoptosis)	+	+
Lung cancer	+	++
Ophthalmological		
Intraocular pressure (reduced)	++	+

*compiled from Russo and Guy [79]

CNS = brain and spinal cord; GI = gastrointestinal

79 Russo and Guy A tale of two cannabinoids 2006

Table A 10. Potential Cannabinoid Benefits to Help Heal 27 Health Problems [80]

Disease and Health Benefits	Cannabinoids					
	CBD	CBC	CBG	CBN	THC	THCV
Analgesia (Reduces pain)	•	•		•	•	
Anti-inflammatory (Reduces inflammation)	•	•	•			•THCA
Anti-insomnia (Sleep aid)	•			•		
Anti-anxiety (Reduces fear, incl.in PTSD)	•					
Anti-depressant (Reduces low energy/fear)					•	
Anti-psychotic (Reduces psychotic behavior)	•					
Anti-spasmotic (Reduces spasms)	•			•		•THCA
Anti-epileptic (Anti seizures)	•					•
Neuro-protective (for neurodegenerative dz)	•				•	
Intestinal anti-prokenetic (digestive aid)	•					
Anti-emetic (reduces nausea)	•				•	
Appetite stimulant					•	
Appetite suppressant						•
Anti-bacterial	•		•			
Anti-microbial (inhibits microorganisms)				•		
Anti-oxidant (fights free radicals in blood)				•	•	
Anti-diabetic (reduces diabetic symptoms)	•					•

80   Based on 74 peer-reviewed published scientific and clinical studies compiled by Cannlabs © 2012; See also leafly.com/knowledge-center

Table A 10 *continued*

Disease and Health Benefits	Cannabinoids					
	CBD	CBC	CBG	CBN	THC	THCV
Anti-psoriatic (helps with itching/psoriasis)	•					
Bone stimulant (helps with bone growth)	•	•	•	•CBDV		•
Anti-proliferative (inhibits tumor growth)	•	•				
Immunosuppressive (helps lupus/RA)	•					
Anti-ischemic (reduces artery blockage risk)	•					
Vasoconstriction (blood vessel constriction)		•				
Vasorelaxant (relaxes veins → ↑ blood flow)	•					
Allograft stimulant (↓ organ rejection)	•					
Intraocular pressure (↓ glaucoma pressure)					•	
Anti-PTSD symptoms (common use reason)	•					

Consider reviewing the '**For Table 10**' paragraph on p. 182 to see how this table came about from the basic science and clinical literature on cannabis. Based on this information **CBD** appears helpful for **20** of these problems and **THC** does for **7**.

Table A 9 suggests this ratio as 14 to 19. A problem is sorting out how to use the rest which include CBC, CBG, CBN and THCV.

We await to see how MM savvy consumers, clinicians and researchers handle all of these bits of information in the context of MM.

In Table A 11 below I summarize what and how temperatures work when heating basic cannabis chemicals.

Effective ratios among all are key (see references).

Delivery and Heat

For Table A 11 – Until recently most users smoked it.

To make delivery safer and work better science and industry creatives found various ways to vaporize cannabis' active components without a regular flame. Vaporizers heat cannabis in a small chamber to temps below its burning point (451°F/233C) which releases effective vapor with the same physical and mental effects as smoking, but without the usual harsh smoke toxins and it smells less. In Table A 11 I summarize what temperatures I found for each of these 9 key cannabinoids that vaporize when heated.

Study it for how you might use these findings.

Table A 11. Vaporization of Cannabinoids & Flavonoids*
Cannabinoid Fahrenheit Celsius Temperature May Help

Cannabinoid	Fahrenheit	Celsius	Temperature	May Help
Water Boils	212	100	Scalds skin	For comparison
Δ1-THC-Acid	220	105	to THC	Euphoriant Analgesic Anti-depressant -inflammatory - emetic -oxidant
Δ9-THC	314	157	Very hot	
Δ8-THC	350	176		
CBD & CBN	320-356	160-180	Hotter	Anxiolytic Analgesic Anti-depressant -psychotic, -inflammatory -oxidant -spasmotic
CBC & THCV	428	220	Even hotter	
CBD-A	248	120		
ß-Caryophyllene	320	160	Hot	Research in progress
Linalool and Humulene	388	198	Hotter	
Your herbs	In Flame	600+ F 316 C	plus toxic tars	

*These temperatures may vary slightly depending on the source.

All these numbers can be confusing, so bear with me and again use trial and error to help sort them out regarding your MM use with any digital device.

Some 'smoke' usually forms at over 360 °F (182 C).

Over 400 F (205 C) the vapor smoke has irritants such as tars, benzene and dioxins. To vaporize, an ideal temperature is said to be from 330 to 375°F (166 to 191 C). THC boils at 392°F (200°C), but active vapors form at about 100 degrees less. [For reference, regular baking foods in an oven spans from slow cooking at 250°F (120 C) to very hot at 450°F (232 C).]

I am unaware of any other drug that requires such detailed heating to both activate and deliver it to a patient. Even when we eat cannabis in food we usually have to heat it to activate it. These methods give us quite an unusual but effective delivery method for a using a "drug" to help heal one or more problems. Without a vaping device that has a digital temperature readout screen the patient has to experiment. Without such a readout it comes back to trial and error again. But it may be useful to know some physical facts about what each cannabinoid and flavonoid tend to become active vapors as a base from which to operate your delivery device or system.

Various cannabinoids are activated between 290-445°F (144-230 C). If you have a digital device, experiment with different temperatures for your personal results.

After all that boring but sometimes useful detail, I and colleagues have tried to make some of the many marketed devices work for the best effect. Warning: some worked well and others, including some costly and hyped ones, did not. Watch if you want some Youtube etc sales pitches, but as our Founders said "Trust but verify," that I call Trial and error. Be sure to have a return *warranty* if it does not deliver for you personally.

Appendix 13 Cannabinoid Basics 4
More Information on *THC*

Throughout this book I have discussed the usefulness of CBD rich cannabis from whole hemp plant extract and its helpful effects as MM. Here, because of its strong effects both plus and minus, I add more details on THC in using it as MM.

THC can be a double edged sword. Depending on important variables, it can impair and/or enhance our mental and physical state, as shown in Table A 12 on the next page. The experience will usually result from a balance of these effects combined with a user's set and setting. Its effect is also determined by the ratio of THC to CBD and the particular strain's other cannabinoids as its synergistic or entourage effect may bring about.

As one of its most practical enhancements, THC in whole plant as MM facilitates our awareness into the present moment, the here and now. Unless our set and setting are negative, in the Eternal Now we will tend to worry less about mundane internal or external conflicts. Used as the whole plant it tends to lift us out of our drama.

Navigating the THC Effect THC appears to open our ordinary intake filter to stimuli such that we experience progressively more of them *all at once*. Without THC, it is usually easier to sort out and focus on something in our outer life and inner life which we may need to address regarding our current well-being, choices and decisions.

The THC effect is *dose related*. A **low dose** user will usually experience less outer and inner life stimuli and effect than that from higher doses. Related to having a finer and less wide-open intake filter, a low dose user tends to have enough ability to function in varying degrees. But the **high dose** user ususally has a more wide-open filter (see also footnote 41 on page 96) and their functioning ability thereby normally lessens. In this fashion, at low doses THC can help us relax and have fun. But higher doses tend to limit our selective attention or ability to focus, as Nadia Solowij and others have shown.

Table A 12. THC's Impairments vs. Enhancements *

Impairments	Comments	Enhancements	Comments
Mental •Altered sense of time •Impaired memory •Inability to focus •Difficulty with acute attention, including to complex information •Difficulty with thinking and problem-solving	All usually increase with THC doses, which often worsen enhancements. Don't make any decisions when effected.	*Mental* •Ability to step out of our personal drama •Keeps us in the mostly positive here & now •Experience expanded consciousness •Some creative insights •↑ Humor and fun.	With safe set & setting. Just enjoy without resulting problems Write down any that may be of interest
Emotional •Fear and anxiety •Numbness	Set and setting always important and worth assuring you are safe. Often when used at high doses or in unsafe set and setting.	Lower THC doses commonly enhance all these; may vary *Emotional* •Empathy for self and others	Needs whole plant parts acting in concert to enhance
Spiritual •Anxiety & Paranoia	As *Emotional* above.	*Spiritual* •Experience connection to Higher Power	Helps connect with safe others
Physical Impairments •Coordination •Balance •Movement •Depth perception •Increased Appetite	Low doses minimize these impairments. Stay where you are in your safe place.	*Physical* Enhancements •Increased relaxation • ↑ Awareness of our 5 senses • May lessen physical &/or emotional pain	With safe set & setting These effects, plus ↑ humor and fun, may be the most common reasons people use

* summarized from Solowij (1998) and Pertwee (2014) and others

Selective attention is one of several processes that together comprise our state of attending to our outer and inner life. In 1890 the psychologist William James described the essence of this process:

"Everyone knows what attention is. It is the taking possession by the mind, in clear and vivid form, of one out of what seem several simultaneously impossible objects or trains of thought. Focalization, concentration of consciousness are of its essence. It implies withdrawal from some things in order to deal effectively with others."

Selective attention allows us to process some stimuli more rapidly and effectively than other mental actions. Harter & Aine (1984) said that it is "the predisposition of an organism to process selectively relevant, as compared to irrelevant, environmental information." Selective attention enhances our processing of relevant stimuli. It is a filtering mechanism protecting our ordinary thinking mind from overload by irrelevant sources of information. As Table A 12 shows, THC *limits* our *selective attention* and *decision making* ability. And so do *other drugs*, from alcohol to benzodiazepines to antidepressants, antipsychotics, "mood stabilizers" and more. But this fact reminds us to limit using cannabis both medically and recreationally, as I caution on page vii in the preface above.

In 1998 Nadia Solowij PhD described how memory impairment is the most consistently reported effect associated with acute cannabis intoxication. She wrote that the most robust explanation for the mechanism of memory impairment is in reduced attention because of increased competition by the intrusion of irrelevant associations. This information supports the observation that THC opens our ordinary intake filter to stimuli such that we experience progressively more of them all at once which makes assessment and decision making about important matters more difficult and risky. This information supports the need during acute intoxication.to write down any new ideas or concerns to reevaluate when no longer affected by THC.

Bibliography

American Cancer Society long statement online: cancer.org/treatment/treatmentsandsideeffects/complementaryandalternativemedicine/herbsvitaminsandminerals/marijuana

Anonymous 1976 *A Course in Miracles*. Course in Miracles Society, Omaha, Nebraska

Backes M 2014 *Cannabis Pharmacy*: The Practical Guide to Medical Marijuana. Black Dog & Leventhal Publishers New York, NY

Bello J 2008 *The Benefits of Marijuana*: Physical, Psychological and Spiritual. Lifeservices Press, Susquehanna, Pennsylvania

Bostwick MJ 2012 *Blurred boundaries*: the therapeutics & politics of medical marijuana. Mayo Clinic Proceedings 87(2) 172-186

Carter GT et al 2004 *Medicinal cannabis*: Rational guidelines for dosing. Drugs 2004 7(5):464-70. See also Mayo Clinic bottom p 193

CBD inhibits FAAH: ncbi.nlm.nih.gov/pmc/articles/PMC3165957/

Chase PL, Pawlik J 1991 *Trees for Healing*: Harmonizing with Nature for Personal Growth and Planetary Balance. Newcastle Publishing

Craven J 2014 *The Power in Pot*: How to Harness the Medicinal Properties of Marijuana in the Management of Clinical and Stress Related Disorders. supportnetstudios.com. Kindle Edition Epub 2005

Crowther SM, Reynolds LA, Tansey EM 2009 The Medicalization of Cannabis. Wellcome Trust Centre, 24 March symposium proceedings

Decorte T, Potter G, Bouchard M 2011 *World Wide Weed*: Global Trends in Cannabis

Cultivation and its Control. Ashgate, Surrey, UK and Burlington, VT, USA

Edwards L 2013 *Awakening Kundalini*: The path to radical freedom. Sounds True, Boulder, CO

Elvick R 2013 Risk of road accident associated with the use of drugs: a systemic review and meta-analysis of evidence. *Accid Anal Prev* 60:254--67

Gardner F 2013 Comes Now Epidiolex FDA approves CBD studies 22 Oct

Gieringer D, St. Laurent J, Goodrich S 2004 Cannabis vaporizer combines efficient delivery of THC with effective suppression *Journal of Cannabis Therapeutics* 4(1) haworthpress.com/web/JCANT

Greyson, CB et al 2007 *Irreducible Mind* – Toward a psychology for the 21st century. Rowman & Littlefield, NY

Hall W, Morely 2015 Possible causes and consequences of reduced perceptions of the risks of using cannabis *Clinical Toxicology* 53, 141–142

Harris B 1990 *Full Circle*: The near-death experience and Beyond. Simon and Schuster Pocket New York, NY

Hergenrather J MD 2014 "Prescribing Cannabis in California" ed. Holland J *The Pot Book*: A Complete Guide to Cannabis, Its role in medicine, politics, science, and culture. Park "Street Press p.416-431.

Herer J *The Emperor Wears No Clothes*. jackherer.com/thebook/ free book to download

Holland J 2010 *The Pot Book*: A Complete Guide to Cannabis Park Street Press, Rochester, VT.

Jiang W et al 2005 Cannabinoids promote embryonic and adult hippocampus neurogenesis and produce anxiolytic- and antidepressant-like effects. *Journal of Clinical Investigation*, 115, 3104-3116

Joy JE et al 1999 Marijuana and Medicine: Assessing the Science Base, Institute of Medicine, March 17, 15 845-1132

Leary T, Metzner R, Alpert R 1964 The Psychedelic Experience: A Manual Based on the Tibetan Book of the Dead. London: Academic Press; University Books. New York, NY

Leonard BE 1997 Therapeutic Uses of Cannabis. British Medical Association Harwood Academic Publishers

Mechoulam R, Parker LA 2013 The endocannabinoid system and the brain. Annual Review of Psychology 64: 21-47 January.
See also Mechoulam R documentary at youtube.com/watch?v=csbJnBKqwIw

Mercola J 2015 *Effortless Healing*. Harmony Books, New York

Pertwee RG ed. 2014 *Handbook of Cannabis* Oxford University Press, Oxford, England – This is the most comprehensive scholarly source for cannabis information that I have found to date

Pertwee RG 2014 Cannabinoid receptor ligands Torcris bioscience. Online at tocris.com/ pdfs/ pdf_downloads/cannabinoid_receptor_ligands_review.pdf

Ring K 1984 *Heading Toward Omega*: In search of the Meaning of the Near-Death Experience. Morrow, NY

Ring K 1998 *Lessons from the Light*: What we Can Learn from the Near-Death Experience. Plenum Press. NY

Rosenthal E 2010 *Marijuana Grower's Handbook*. Quick American Archives; Special Edition

Russo E, Guy GW 2006 A tale of two cannabinoids: The therapeutic rationale for combining tetrahydrocannabinol and cannabidiol *Medical Hypotheses* 66:234–246. Online site: intl.elsevierhealth.com/journals/mehy

Russo EB 2011 Taming THC: potential cannabis synergy and phytocannabinoid-terpenoid entourage effects *Br J Pharmacol*. 163(7): 1344–64.

Sannella L 1987 *The Kundalini Experience, Psychosis or Transcendence?* Integral Publishing.

Skeptics on cannabis 1: sciencebasedmedicine.org/medical-marijuana-as-the-new-herbalism-part-1-the-politics-of-weed-versus-science/

Skeptics on cannabis 2: sciencebasedmedicine.org/medical-marijuana-are-we-ready/

Solowij N 1998 *Cannabis and Cognitive Functioning*. Cambridge U Press, NY

Solowij N, Yucel M, Lorenzetti V, Lubman D 2012 Does cannabis cause lasting brain damage? in *Marijuana and Madness* 2nd ed D Castle, Cambridge University Press, NY

Tart C 1971 *On Being Stoned*: A psychological study of marijuana intoxication. Science and Behavior Books

Thornton L 2014 Through Heaven's Gate and Back. Lulu

Umathe SN et al 2009 Endocannabinoids mediate anxiolytic-like effect of acetaminophen via CB1 receptors. *Prog Neuropsychopharmacol Biol Psychiatry*. 2009 Oct 1;33(7):1191-9

Ware MA, Adams H, Guy GW 2005 The medicinal use of cannabis in the UK: results of a nationwide survey. *Int J Clin Pract*. Mar;59(3):291-5

Werner, Clint (2012-02-11). *Marijuana Gateway to Health*: How Cannabis Protects Us from Cancer and Alzheimer's Disease (Kindle Location 2). Dachstar Press. Kindle Edition Has a good table showing CBD/THC ratios among 17 selected strains

Whitfield BH 1995 *Spiritual Awakenings*: Insights of the Near Death Experience and other doorways to our Soul. Health Communications Inc. Deerfield Beach, FL

Whitfield BH 1998 *Final Passage*: Sharing the journey as this life ends. Health Communications Inc. Deerfield Beach, FL

Whitfield, BH 2009 *The Natural Soul:* Unity with the Spiritual Energy that Connects Us. Muse House Press, Atlanta, GA

Whitfield BH 2011 *Victim to Survivor and Thriver*. Muse House Press, Atlanta, GA

Whitfield BH, Cormier S, 2013. *AFGEs*: A Guide to Self-Awareness and Change. Muse House Press, Atlanta, GA

Whitfield BH 2009 Mental and emotional health in the kundalini process. in *Kundalini Rising*: Exploring the energy of awakening. Sounds True, Boulder CO

Whitfield CL 2009 Spiritual energy: perspectives from a map of the psyche and the kundalini recovery process. in *Kundalini Rising*: Exploring the energy of awakening. Sounds True, Boulder CO

Whitfield CL 1998 *Choosing God*: A Bird's-Eye-View of A Course in Miracles. Muse House Press, Atlanta, GA

Whitfield CL1987 *Healing the Child Within*: Discovery & Recovery for Adult Children of Dysfunctional Families. Health Coms, Deerfield B, FL

Whitfield CL: *A Gift to Myself*. Health Communications, Deerfield B, FL 1990

Whitfield CL 1993 *Boundaries and Relationships:* Knowing, Protecting and Enjoying the Self. Health Communications, Deerfield Beach, FL

Whitfield CL 2003 *The Truth about Depression*: Choices for healing. Health Communications, Deerfield Beach, FL

Whitfield CL, 2004 *The Truth about Mental Illness*: Choices for healing. Health Communications, Deerfield Beach, FL.

Whitfield CL 2004 *My Recovery:* A personal plan for healing. Health Communications, Deerfield Beach, FL

Whitfield CL, Whitfield BH, Jyoti, Park R, 2006 *The Power of Humility*: Choosing Peace over Conflict in Relationships. Health Communications, FL

Whitfield CL 2012 *Not Crazy*: You May NOT be Mentally Ill . Muse House Press, Atlanta, GA

Wolff J. 2010 "Thots on Pot", Chapter 35 from *The Pot Book* Holland J ed. Park Street Press, Rochester, VT

Late Referencess added: Gable RS 2004 Comparison of acute lethal toxicity of commonly abused psychoactive substances. *Addiction* 99: 686 – 696 Added late to references

Gallily R, Yekhtin Z, Hanus LO 2015 Overcoming the Bell-Shaped Dose-Response of cannabidiol by using cannabis extract enriched in cannabidiol. Pharmacology & Pharmacy 6:75-85 Added late

Zimmer L, Morgan JP 1998 Marijuana Myths and Marijuana Facts: A Review of the Scientific Evidence. British Medical Association. See also John's talk on stopthedrugwar.org/chronicle/2008/feb/21/ [add next text here] memoriam_dr_john_p_morgan

Zinberg NE 1984 *Drug, Set, and Setting*: The Basis for Controlled Intoxicant Use. Yale University Press

Other References - from the Internet

Erowid erowid.org/plants/cannabis/

Learn more here: theweedblog.com/

Herer J, his free book to download. *The Emperor Wears No Clothes*. Online at jackherer.com/thebook/

The Future Of Marijuana In Clinical Practice: Q&A With
Dr. Sunil Aggarwal

leafscience.com/2013/09/24/interview-dr-sunil-aggarwal-future-marijuanlinical-practice/

An 18 minute video (from 2011) telling my near-death experience and the research I did after at a University Medical School youtube.com/watch?v=r-RCxZIv3Td8 [A must if interested in NDEs]

A 7 1/2 minute video from 1989 of my near-death experience youtube.com/watch?v=gF_Dj6EduLY

Marijuana Documentary April 20, 2014– Cannabis Research Studies youtu.be/sYA9EpVB2qo

The Consciousness of Trees. Video
expandedconsciousness.com/2014/03/07/the-consciousness-of-trees-video/
Cannabis research A to Z: calgarycmmc.com/ [may have to register first to view large content]

Rappaport T, Leonard-Johnson S 2014 *CBD-Rich Hemp Oil*: Cannabis Medicine is Back. Cannabismedicineisback.com

Howlett AC, Barth F, Bonner TI, Cabral G, Casellas P, Devane WA, et al 2002 Classification of cannabinoid receptors. *Pharm Rev* 54:161–202

mayoclinic.org/drugs-supplements/marijuana/evidence/hrb-20059701

Armentano P, Carter GT, Sulak D, Goldstein ET et al 2013 Emerging Clinical Applications For Cannabis & Cannabinoids: A Review of the Recent Scientific Literature, 2000-13 **Excellent summary** update at norml.org/pdf_files/NORML_Clinical_Applications_for_Cannabis_and_Cannabinoids.pdf

More MM Information, News and Updates

Glossary for CBD related terms is at projectcbd.org/medicine/glossary/

Drugwarfacts.org

Death by 'drugwar' pages: drugwarrant.com/articles/drug-war-victim/Drug

Cannabis interactions calgarycmmc.com/interactions.htm Leafly.com

Marijuana Policy Project MPP.org

Medicalmarijuana.procon.org/ website with and news on MM

Medicaljane.com Useful information, news and links on MM

Naturalmedicines.therapeuticresearch.com/ - for dosing information

Projectcbd.org/ by Fred Gardner and Martin Lee

medicalcannabis.net/cannabis-as-medicine/

medicalcannabis.comproplantware.com/index.htm Nate and Emily Morris

Website example how-should-we-classify-groupings-ofcbdthc-ratios

Also search online for CBD THC ratios in using MM

Safeaccessnow.org

Stopthedrugwar.org

Tokeofthetown.com/

themedicalcannabisinstitute.org and medicalcannabis.com/

ukcia.org/ UK Cannabis Internet Activist – The cannabis information site

US Patents appl calgarycmmc.com/canna bispatents.htm vape-nation.com

vaporize-marijuana-guide/ & their guide-to-cannabinoids

vaporizerblog.com/faq/ steephilllab.com/resources/cannabinoid-and-terpenoid-reference-guide/

kindgreenbuds.com/medical-marijuana/cannabidiol-a-cannabis-sativa-constituent-as-an-antipsychotic-drug/

CBD short video on vimeo.com/58945962

Dateline nbcnews.com/nightly-news/video/parents-fight-to-make-medical-marijuana-legal-for-sick-child-459364931854

Harvey B, Bobroff M et al 2014 *The Culture High.* One of the best documentaries ever made. Do not miss it. Take notes.

INDEX

About the Author

Barbara Harris is a consciousness researcher and integrative counselor in private practice in Atlanta, working with people who have had repeated childhood trauma, co-dependence, addictions and problems in living. She is also a Near-Death Experiencer, and works with others who have had similar Spiritually Transformative Experiences. She started out as an ICU respiratory therapist in the early 1980s. Soon she was teaching at respiratory and nursing conferences and writing papers for their journals on a new topic that she called "The Emotional Needs of Critical Care Patients." Her background was as a respiratory therapist and certified massage therapist.

Barbara was a prime subject in Kenneth Ring's seminal book Heading Toward Omega: In search of the meaning of the Near-Death Experience. He wrote about her again in Lessons from the Light: What we can Learn from the Near-Death Experience. She then became a researcher at the University of Connecticut Medical School assisting psychiatry professor Bruce Greyson Director of Research for the International Association for Near-Death Studies (IANDS.) She went on to teach at Rutgers University's Institute for Alcohol and Drug Studies for 12 years, calling her classes "When the 12th Step happens first," based on her research into the aftereffects of spiritually transformative experiences.

She then went back to school to study Holistic Health focusing on the Eastern models of the energy of the human body and combined this knowledge with her respiratory training to develop new techniques for breath and energy work as a way to help release blocks in the bioenergy system. She developed this breath and energy/body work method with patients while working with Alan Gaby, MD, a founder of nutritional medicine and past president of the American Holistic Medical Association.

Barbara served on the board and executive board of the International Association for Near-Death Studies (IANDS).

She is past president and board member of the Kundalini Research Network. Barbara taught classes in 2012 for the first annual ACISTE conference (American Center for the Integration of Spiritually Transformative Experiences) on *The Power of Humility*: A Map of the Integration Process (from her co-authored book The Power of Humility 2006. Health Communications, Inc. Deerfield Beach, FL).

She's been an expert guest on major television shows. Her research and personal story are in documentaries in Japan, England, Belgium, France, Germany, Canada and the US. She has given talks for the Senate on Capitol Hill and The United Nations in New York.

Barbara has authored five self-help books on near-death experiences, death and dying and consciousness research (see bibliography). She co-authored six more. She has written numerous chapters for other books and co-authored papers for scholarly psychology and consciousness research journals.

Other Books from our library